Nuclear Is Not the Solution

The Folly of Atomic Power in the Age of Climate Change

M. V. Ramana

VERSO

London • New York

First published by Verso 2024
© M. V. Ramana 2024

1 3 5 7 9 10 8 6 4 2

Verso
UK: 6 Meard Street, London W1F 0EG
US: 388 Atlantic Avenue, Brooklyn, NY 11217
versobooks.com

Verso is the imprint of New Left Books

ISBN-13: 978-1-80429-000-2
ISBN-13: 978-1-80429-001-9 (UK EBK)
ISBN-13: 978-1-80429-002-6 (US EBK)

British Library Cataloguing in Publication Data
A catalogue record for this book is available from the British Library

Library of Congress Cataloging-in-Publication Data

Names: Ramana, M. V., 1966- author.
Title: Nuclear is not the solution : atomic power in the age of climate
change / M. V. Ramana.
Description: London ; New York : Verso, 2024. | Includes bibliographical
references and index.
Identifiers: LCCN 2024009756 (print) | LCCN 2024009757 (ebook) | ISBN
9781804290002 (hardback) | ISBN 9781804290019 (UK ebook) | ISBN
9781804290026 (US ebook)
Subjects: LCSH: Nuclear power plants—Risk assessment. | Nuclear policy. |
Climate change mitigation—Government policy.
Classification: LCC TK9152.16 .R36 2024 (print) | LCC TK9152.16 (ebook) |
DDC 363.7/06—dc23/eng/20240319
LC record available at https://lccn.loc.gov/2024009756
LC ebook record available at https://lccn.loc.gov/2024009757

Typeset in Sabon by MJ & N Gavan, Truro, Cornwall
Printed and bound by CPI Group (UK) Ltd, Croydon CR0 4YY

Contents

Acknowledgments

This book has taken a very long time to evolve from conception to fruition, although not as long as it takes to build a nuclear plant. Over this journey, I have benefited from numerous people, and it would be impossible to name all of them. Below is just a short list.

I will start with Peter Dimock and Andy Lichterman. Conversations with them underlie much of the book—in Andy's case, too numerous to even try to count. Both had their own visions for what this book should have been. While I found myself unequal to fulfilling their ideas, I have still gained enormously from those discussions and am indebted to them for their generosity.

The book was also greatly shaped by Nicole Aschoff, who made me change its focus and scope, and who helped me get a contract with Verso Books. This book might have been much less compelling but for Nicole's interventions, and for that I am deeply obliged. Asher Dupuy-Spencer took over when Nicole had to leave and left his mark on the manuscript in many ways, too, shepherding this book in its journey to publication, interfacing with the many people at Verso and Penguin Random House whose labor was essential to its production. I am grateful to both of them for their patience with me and my inability to meet deadlines. As Peter Dimock emailed me in May 2019, "books are always very jealous of their author's attentions and tend to vanish at the first hints of life's more pressing (and probably more important?) embodied obligations." I can only hope that the other obligations I encountered in the last many years were pressing and important enough to justify the book frequently vanishing from my work life. Also, I should thank freelancer Conor O'Brien and Jeanne Tao at Verso, whose edits made this into a much better book than it might have been otherwise.

Numerous friends pitched in to read drafts of sections of the book and offer comments, criticisms, and suggestions. Both the book and I have profited from their feedback and subsequent exchanges. There are too many of these people for me to be sure that I have listed all. But among them, four names stand out: Bob Jensen, CC Huang, Cassandra Jeffery, and John Downer. All four read through practically all of my chapters and guided me throughout this journey. I am forever indebted to them for their kindness and generosity.

The writing group in the 2020 Wall Institute cohort deserves gratitude for moving me from just talking about writing a book to producing something that could be shared with others, including a proposal that I could send out to potential publishers. My sincere appreciation to Jennifer Black, Carrie Jenkins, Michelle Stack, Shiela Teves, and above all, the late Y-Dang Troeung. Y-Dang metaphorically held my hand and helped me write in ways that were not within my comfort zone, and I dearly wish she were alive to see this book in print.

Other readers that I am grateful to are Bulgan Batdorj, Friederike Frieß, Jixiang Wang, Johanna Höffken, Sara Nelson, Sria Chatterjee, Susan O'Donnell, and Thomas Bass. All of them took the time to not just comment on my drafts but also encourage me to keep improving what I was writing. And then there are other friends who did not read any of these chapters but have shaped this book's contours in one way or the other. I gratefully acknowledge conversations with Achin Vanaik, Allison Macfarlane, Ansar Fayyazuddin, Aparna Lakshmi, Ed Lyman, Frank von Hippel, Gordon Edwards, Harini Rajagopal, Jahnavi Phalkey, Jean Kumagai, John Byrne, Juliet Lu, Kajoli Banerjee Krishnan, Kavita Philip, Linda Pentz Gunter, Maya

Wind, Naomi Klein, Nagraj Adve, Ray Acheson, Raza Mir, Sadia Tasleem, Sajan Saini, Samanth Subramanian, Sharach-chandra Lele, Sharon Weiner, Sophie Gaur, Stephanie Cooke, Susan D'Agostino, and Tara Cookson. Finally, there is the larger community of friends that has supported me through this process—I won't even try to name them.

This book is based in part on articles written with a number of collaborators, including Aileen Murphy, Alexander Glaser, Ali Ahmad, Amory Lovins, Amy King, Arjun Makhijani, Andrei Covatariu, Ashwin K. Seshadri, Athena Kerins, Avino Niphi, Benjamin Sovacool, Bernadette Cogswell, Cassandra Jeffery, Cathy Kunkel, Eri Saikawa, Friederike Frieß, Friga Siera, Isha Bhasin, Johanna Höffken, Jungmin Kang, Kerrie Blaise, Laura Berzak Hopkins, Lindah Ddamba, Matthias Englert, Nataliawati Siahaan, Priscilla Boakye Atansah (Agyapong), Rajendran Pillai, Richard Tanter, Sara Nelson, Shakiba Fadaie, Stephen Thomas, Suvrat Raju, and Zia Mian. I thank all of them for working with me and sharing their insights.

I also appreciate the numerous journals, magazines, and newspapers that have published my articles. Among them, the *Bulletin of the Atomic Scientists* and its editor John Mecklin and *Against the Current* and its editor Dianne Feeley deserve special mention for allowing me the space to articulate some of my arguments about the political economy of nuclear energy.

Mycle Schneider and Antony Froggatt, along with Julie Hazemann and Nina Schneider, have been bringing out the World Nuclear Industry Status Report, a remarkable compilation of data, year after year, against all financial odds. It has been my privilege to be part of this team, and to draw on these reports in this book.

My position at the University of British Columbia owes itself, in part, to the commitment of Jennifer Simons to financially support work on nuclear disarmament and related topics in Canada. I warmly acknowledge her generosity. Earlier, at Princeton University, I was supported by a range of foundations and I am grateful to them as well. I also greatly appreciate Scott Denman's support for my public interventions on the topic of nuclear energy.

I am beholden to the two institutional homes where most of this book was conceptualized or written, and where I undertook the research underpinning it. These are, chronologically, the Program on Science and Global Security at Princeton University and the School of Public Policy and Global Affairs at the University of British Columbia in Vancouver. I deeply appreciate the support of all my coworkers and colleagues, too numerous to name here, at both institutions.

Finally, my dear family, which has put up with my nuclear obsessions, frequently with humor, occasionally with puzzlement, and always with affection. Over the period I worked on this book, my daughters Shruti and Swara have grown into feisty young women, full of their individual ideas about the world and their own commitments to change it for the better. I can only hope that my efforts help.

I dedicate this book to Geetha, without whom it could not have been written, and to Bubbles, who curled up close to me during many of the days and nights I worked on it.

This book contains hundreds of endnotes, all in invisible ink. They had to vanish to make this into a slimmer volume that you can hold more easily. For those interested in the sources for my statements and arguments, books that influenced my work in some way, and just in reading more, please look at https://sppga.ubc.ca/nuclear-is-not-the-solution.

Introduction

I am scared about how fast climate change is disrupting our world. At a theoretical level, I have known for decades about growing carbon dioxide emissions and resultant changes to global and local temperatures, sea-level rise, severe storms, wildfires, and so on. But it was not till 2012, when Hurricane Sandy hit the northeast of the United States, that I was directly impacted. The power of that storm was immense, but I knew—theoretically, of course—that people elsewhere had experienced far worse storms.

More recently—in August 2023, as I was finishing this book—it was the turn of wildfires. As the McDougall Creek wildfire came closer to the University of British Columbia (UBC) campus in Kelowna, students and staff were asked to evacuate. My daughter Shruti is a student there. Because it was summer, she was at home in the Vancouver campus of UBC, where I teach. Had the fires occurred just two weeks later, I would have definitely been panicking.

I can go on for much longer in this vein. But there isn't any need. Just about anyone alive today has been impacted in some way by climate change. Others have written at length about how the climate crisis is intensifying by the year, and one can stock a small library with published books about the myriad risks flowing from climate change. The library would be even larger if one included literature on the other related multiple cascading ecological crises we are confronting.

As someone trained in physics, and as an academic paid to research, I have been drawn to studying one essential

contributor to these crises: how energy and electricity are produced, especially those methods proposed to mitigate climate
change. Prominent among these proposals is nuclear energy.
Although climate change scares me, I am even more scared of
a future with more nuclear plants. Increasing how much energy
is produced with nuclear reactors would greatly exacerbate the
risk of severe accidents like the one at Chernobyl, expand how
much of our environment is contaminated with radioactive
wastes that remain hazardous for millennia, and last but not
least, make catastrophic nuclear war more likely.

Some might argue that these risks are the price we must pay
to counter the threat of climate change. I disagree, but even
if one were to adopt this position, my research shows that
nuclear energy is just not a feasible solution to climate change.
A nuclear power plant is a really expensive way to produce
electricity. And nuclear energy simply cannot be scaled fast
enough to match the rate at which the world needs to lower
carbon emissions to stay under 1.5 degrees Celsius, or even
2 degrees.

Cost and the slow rate of deployment largely explain why
the share of global electricity produced by nuclear reactors
has been steadily declining, from around 16.9 percent in 1997,
when the Kyoto Protocol was signed, to 9.2 percent in 2022.[1]
In contrast, as the costs of wind and solar energy declined
dramatically, and modern renewables (which do not include
large dams) went from supplying 1.2 percent of the world's
electricity in 1997 to 14.4 percent in 2022.

Another contrast is revealing. When pro-nuclear advocates
talk about solving climate change with nuclear energy, they
call for building lots and lots of reactors. The World Nuclear
Association, for example, proposes building thousands of
nuclear reactors, which would together be capable of generating a million megawatts of electricity, by 2050. Such a goal is
completely at odds with historical rates of building nuclear
reactors.[2]

Some proponents of nuclear energy refuse to give up on the technology. They blame the decline in nuclear energy and the high costs and long construction periods on the characteristics of older reactor designs, arguing that alternative designs will rescue nuclear energy from its woes. In recent years, the alternatives most often advertised are small modular (nuclear) reactors—SMRs for short. These are designed to generate between 10 and 300 megawatts of power, much less than the 1,000–1,600 megawatts that reactors being built today are designed to produce.

For over a decade now, many of my colleagues and I have consistently explained why these reactors would not be commercially viable and why they would never resolve the undesirable consequences of building nuclear power plants (more on this topic in chapter 6). I first started examining small modular reactors when I worked at Princeton University's Program on Science and Global Security. Our group largely comprised physicists, and we used a mixture of technical assessments, mathematical techniques, and social-science-based methods to study various problems associated with these technologies. My colleague Alex Glaser, for example, used neutronics models to calculate how much uranium would be required as fuel for SMRs, which we then used to estimate the increased risk of nuclear weapons proliferation from deploying such reactors.[3] Zia Mian, originally from Pakistan, and I showed why the technical characteristics of SMRs would not allow for simultaneously solving the four key problems identified with nuclear power: its high costs, its accident risks, the difficulty of dealing with radioactive waste, and its linkage with the capacity to make nuclear weapons.[4] My colleagues and I also undertook case studies on Jordan, Ghana, and Indonesia, three countries advertised by SMR vendors as potential customers, and showed that despite much talk, none of them were investing in SMRs, because of various country-specific reasons such as public opposition and institutional interests.[5]

We were not the only people coming up with reasons for not believing in the claim that new reactor designs would solve all these problems. Other scientists and analysts also highlighted the dangers and false promises of SMRs.[6]

Nuclear advocates are not deterred by such arguments. They insist that this time it will be different. Nuclear plants would be cheap, would be quick to build, would be safe, would never have to be shut down in unplanned ways, and would not be affected by climate-related extreme weather events. The evidence from the real world, which I elaborate on later, suggests otherwise. Nuclear reactors are unlikely to possess any of these characteristics, let alone all of them. Thus, what is actually being advocated might be termed *faux* nuclear plants, existing only in the imagination of some, not in the real world.

My bottom line is that nuclear energy, whether with old reactor designs or new *faux* alternatives, will simply not resolve the climate crisis. The threat from climate change is urgent. The world has neither the financial resources nor the luxury of time to expand nuclear power. Meanwhile, even a limited expansion would aggravate a range of environmental and ecological risks. Further, nuclear energy is deeply imbricated in creating the conditions for nuclear annihilation. Expanding nuclear power would leave us in the worst of both worlds.

Too virtuous to meter?

Proponents of nuclear energy have other reasons to support their preferred technology. They argue that nuclear reactors can do much more than just generate electricity. The "much more" depends on the specific context, and could include creating well-paying jobs, boosting national pride, providing energy independence, supplying clean water, and producing medical isotopes to treat cancer. As the public has become more concerned about climate change, nuclear advocates have appended to this list two more applications for energy from nuclear reactors: capturing carbon dioxide from the atmosphere (direct air capture) and producing hydrogen and high-temperature heat for industrial processes.

All of these are reminiscent of what Admiral Lewis Strauss, one of the central characters in the hit Hollywood film *Oppenheimer* and the chair of the US Atomic Energy Commission in the 1950s, told the National Association of Science Writers on September 16, 1954. Ten days after the groundbreaking for first US nuclear plant, Strauss told his audience that given the great promise of nuclear technology, it would not be "too much to expect that our children will enjoy in their homes electrical energy too cheap to meter."

The many claims about what else nuclear reactors can do make one wonder: Is nuclear energy too virtuous to meter?

Let me offer one example from a company called Hyperion Power Generation offering a small nuclear power plant design that was actively covered in the media between 2007 and 2012. In March 2010, the founder of this company, John Deal, told the *Albuquerque Journal*, "We started this company to clean water in Africa ... Our emphasis is helping people not die from not having clean water ... If you've got energy, you can have all the clean water you want."

This was not a one-off sales pitch. In their 2011 article in *Issues in Science and Technology*, writer Ross Carper and academic Sonja Schmid offer this description of Deal in action:

> In the middle of Deal's talk in Denver, he began flipping through some artist-drawn images. The most striking of all shows a small nuclear reactor, buried and unattended at what looked to be less than 15 feet below the surface. Two simple tubes snake upward from the reactor, drawing the eye to a pair of gray above-ground tanks, with the words "Potable Water" stamped on the side. The setting? An impoverished African village complete with

about a dozen mud-constructed, thatch-roofed huts. A handful of people were drawn into the image, all of them walking to or from the clean water source, which is apparently powered by a $50 million HPM.[7]

HPM stands for Hyperion Power Module, the nuclear reactor the company was advertising, and the cost estimate of $50 million for a nuclear reactor should be seen in that light—as wishfully cheap. (A few years later, PitchBook, a database of private equity-based corporations, listed the company as "out of business.")

Such promises of atomic energy delivering progress to Africa date back to the beginning of the nuclear age. On January 28, 1947, for example, Waldemar Kaempffert, the science editor of the *New York Times*, predicted,

> The desert of Sahara could easily be irrigated by electric pumps driven by uranium power, with the result that more surplus cotton than we could sell at a profit and more surplus plant food than we could eat would be dumped on the market. Africa would be transformed into another Europe, with savages [*sic!*] who never saw a steam shovel or railway train transformed into machine tenders.[8]

After more than half a century of experience with nuclear technology, ideas about using it to provide clean water to poor people are delusional at worst and deceptively self-serving at best. Reducing the problem of insufficient clean water to an absence of energy ignores the many other problems that prevent African villagers from accessing clean water and the persisting legacies of colonialism and imperialism that led to "underdevelopment" in the first place.[9]

In his "communal memoir" of the aerospace industry *Blue Sky Dream*, the journalist David Beers talks about a special characteristic of the former Nazi rocket scientist Wernher von Braun, the man sometimes termed "the father of America's space program" due to his important role in transferring rocket technology to the United States.

The classic American entrepreneurial hero searches out unmet desires in the everyday world and then, with a certain flexible flair, invents the answers, products for the masses to use. Von Braun's genius lay elsewhere. He was brilliant at inventing new and different uses for the only product he ever desired to make, the space rocket. He was a master at selling his one product to the only customers who could ever afford it, a nation's rulers.[10]

Much like von Braun, vendors and advocates of nuclear power are really interested only in selling nuclear reactors, and they try to invent different uses for their favored product. Delivering clean water, heating houses or industries, and propelling rockets and ships are all only vehicles for selling nuclear reactors. However, the appeal to other uses for nuclear reactors is also, simultaneously, an expression of the inability of the technology to economically deliver on its primary product: electricity. It is the weakness of the nuclear industry that forces it to seek alliances with other constituencies.

Too destructive to meter?

Nuclear energy does have one virtue, but it is one that its advocates, for the most part, avoid mentioning: its innate and inseparable connection to nuclear weapons, and more generally, to the military. I use the word "virtue" to mean both an inherent attribute and an asset beneficial to its proponents. Technically, there are significant overlaps between the apparatus needed to produce nuclear energy and what is needed to produce the fissile material, the hardest step in acquiring nuclear weapons. In addition, personnel can be interchanged between the nuclear energy and weapons programs. And finally, there are institutional incentives for organizations developing nuclear energy to get involved in making nuclear weapons, due to the political power that flows from the latter. Nuclear technology also contributes to powering long-range submarines, especially those used to fire off nuclear missiles, and to providing the material to manufacture depleted uranium munitions used in Iraq and Ukraine. I elaborate on these connections in chapter 5.

Nuclear energy advocates often argue against conflating nuclear energy with nuclear weapons, but the connection is visible for all those who want to look. As of September 2023, 275 of the 410 nuclear reactors labeled as operating by the International Atomic Energy Agency are in countries possessing nuclear weapons. Add countries like Canada and Japan that are militarily allied with nuclear weapon states, and the overlap is staggering. While it is certainly true that not all countries with nuclear energy have produced nuclear weapons, they are closer to being able to do so than they would be if they had never built nuclear reactors.

The overlap between the two technologies was obvious to most knowledgeable people at the beginning of the atomic age. In 1946, when discussing a proposal for the international control of nuclear weapons, Robert Oppenheimer, the head of the program that produced the first atomic bombs, which destroyed Hiroshima and Nagasaki, expressed it thus: "We know very well what we would do if we signed such a convention: we would not make atomic weapons, at least not to start with, but we would build enormous plants, and we would design these plants in such a way that they could be converted with the maximum ease and the minimum time delay to the production of atomic weapons."

Within a few years, however, countries with nuclear technology started a sustained campaign to get the public to think differently about nuclear energy, most notably after President Dwight Eisenhower's "Atoms for Peace" speech in 1953. This "greatest of destructive forces," Eisenhower prophesized, "can be developed into a great boon, for the benefit of all mankind," can be put to "universal, efficient and economic usage" and whose "special purpose would be to provide abundant electrical energy in the power-starved areas of the world."

In other words, forget the destructive capacity of nuclear energy. Just focus on what a wondrous future it can create. The Soviet counterpart of this effort is captured by the slogan "May the atom be a worker, not a soldier." The hope seems to be that by pretending that nuclear energy was not linked to weapons, public fears about the destruction that would result from the use of nuclear weapons would be quelled.

Institutions and governments around the world developing nuclear technology often start by touting its potential to produce electricity. This was the case in India. For over two decades, India's Atomic Energy Commission was ostensibly working on nuclear energy only "for peaceful purposes," until the 1974 test of a nuclear weapon blew up that pretense.[11]

Many private companies profit enormously from both nuclear energy and nuclear weapons. Examples include Bechtel, Babcock & Wilcox (now BWX Technologies), and Fluor in the United States, Larsen & Toubro in India, and Rolls Royce in the United Kingdom. While there might not be a similar level of involvement by private companies in countries like China, where public sector and national organizations play the analogous roles, the differences between the two categories are not very material to understanding the structure of, and trends in, the nuclear sector. National laboratories contract out work and are sometimes even managed by private companies. And private companies thrive on public contracts that they often have exclusive access to, belying any notion of free markets and competitive entrepreneurship.

For both corporate and governmental entities, nuclear technology is a wonderful asset. As analyst and disarmament activist Andrew Lichterman argues:

The nuclear road provides elites in nuclear establishments with privileged access to their own country's resources, a development context that can be shielded from foreign competition, and forms of trade and industry that can be portrayed as increasing in importance as fossil fuels diminish. This is so whether the intention to develop nuclear weapons is clear or is allowed to remain ambiguous. The powerful tools of nationalism and 'national security' secrecy can be used to facilitate the extraction of wealth from the rest of society and prevent scrutiny of national nuclear enterprises that whether in first generation nuclear powers or post-colonial states have been rife with technical problems, corruption, and widespread, intractable environmental impacts.[12]

Overview of the book

The chapters that follow explain why expanding nuclear power production is neither a desirable nor a feasible solution to climate change. Due to the use and production of radioactive materials at reactors, expanding nuclear energy to mitigate climate change will inevitably result in a variety of undesirable risks and environmental impacts. Nor is it compatible with environmental and social justice.[13] The consequences and burdens of such an expansion will fall primarily on communities that are distant from the centers of power, and economically and politically too marginal to figure in the calculations of decision makers.

In chapter 1, I explain how all nuclear reactors, including small ones, are at risk for severe accidents due to their intrinsic technological characteristics. When it comes to nuclear facilities, I will argue, there is nothing that fits a strict definition of "safe." The risk is exacerbated by a range of factors, including extreme weather patterns due to climate change, the multiple and conflicting priorities of organizations operating nuclear facilities, and the weakening of regulation by industry lobbyists and other powerful economic actors. Accidents, when they occur, produce radioactive contamination that reaches across space and time; thirty-five years after the Chernobyl accident, parts of Ukraine and Belarus are still uninhabitable because of high radiation levels. Radioactive cesium released by the disaster was found in sheep in England, which remained contaminated for decades; restrictions on eating these sheep were lifted in all areas only in 2012.

Expanding nuclear energy production will also result in a growing inventory of radioactive wastes, no matter what kinds of reactors are used. Some of these wastes remain radioactive, and thus hazardous to human health, for hundreds of thousands of years. Despite decades of well-funded research, there is no demonstrated way to safely manage them, and because of the long periods involved, there will always be uncertainties about the fate of these materials.[14] As a result, it is likely that radioactive materials will contaminate the biosphere at some point in the future. This is an important cause for opposition from communities near sites chosen for nuclear waste repositories.

Another concomitant activity to the operation of reactors is uranium mining, which has been responsible for contaminating land and water around the world, especially in areas occupied by Indigenous communities. Given these inevitable impacts, nuclear power is neither clean nor sustainable.

One way that some nuclear energy advocates try to get around these conclusions is by claiming that exposure to radiation is harmless, at least below some threshold. But as I explain, there is ample evidence that exposure to radiation, even at low levels, leads to cancers and other negative health outcomes.

In chapter 2, I argue that nuclear energy is not a feasible solution to climate change. The potential role of any energy technology in climate change mitigation depends on how costly it is and how fast it can displace fossil fuels. Nuclear power is today among the costliest ways of generating electricity. In the United States, home to the most nuclear plants globally, the financial firm Lazard estimated in October 2021, before the war on Ukraine created supply chain–related uncertainties, that power from a new nuclear plant costs over four times the corresponding costs of power from wind turbines and utility-scale solar plants respectively. Annual estimates from Lazard and other sources also show that the costs of renewables and energy storage are going down rapidly, whereas nuclear construction costs have been rising.

In addition, utilities in many countries are shutting down older nuclear reactors because their operating costs are too high to compete in electricity markets. The alternative to this has been to appeal for and obtain government subsidies, usually on the backs of ratepayers. Nuclear power, then, is not thriving but barely surviving on life support in many countries.

The second factor to be considered when evaluating whether an energy technology can help with climate change is how much time it would take to expand. The Intergovernmental Panel on Climate Change and other international bodies have warned that to stop irreversible damage from climate change, emissions have to be reduced drastically by 2030. Nuclear energy cannot even begin to contribute within that time frame. Measured from when construction commences, a nuclear plant takes at least a decade to start producing electricity. Planning, regulatory evaluations of new designs, and raising finances might add another decade.

The result of these economic and temporal realities is that nuclear power capacity has largely been stagnant, with a falling share of electricity production; this trend will continue into the future. Thus, I conclude, even if one were to ignore the undesirable outcomes associated with nuclear energy, the technology cannot meaningfully contribute to mitigating climate change. Pursuing such false solutions only lowers the chances of dealing with the environmental problems at hand.

Chapters 1 and 2 lead me to the question: Despite these intractable problems, why do governments and private corporations continue to fund nuclear power? The rest of the book explores some reasons for this support, in the process making a novel argument about why the political economy of nuclear energy constitutes a separate reason to dismiss it as a viable solution to climate change.

In chapter 3, I show how several large and financially powerful organizations have profited from building and operating nuclear plants by making the public pay for the high costs of these activities, through either electricity bills or taxes. The high costs, then, become assets that allow for greater profits. Even in the event of a severe accident, owners of nuclear plants pass on their liability to the public. In the United States, this transfer happens through the Price-Anderson Act, first passed in 1957. Legislations such as the Price-Anderson Act cap the financial liability to nuclear plant owners at levels far lower than the actual costs of dealing with severe accidents like the one at Fukushima—roughly a tenth of Japan's GDP in 2019.[15] The government—in other words, the taxpayer—pays the rest. The public also will have to pay the long-term expenses associated with dealing with the multiple forms of radioactive waste and the subsidies aimed at inducing private companies to invest in nuclear power.

Such practices that benefit the wealthy at the cost of the many are also found in other sectors of the economy. As Noam Chomsky explained in an interview about the financial crash of 2008, "What you have is a system of socialization of cost and risk and privatization of profit. And that's not just in the financial system. It is the whole advanced economy." This systemic feature, which allows many powerful organizations to benefit economically from the nuclear energy business, underlies many decisions concerning nuclear power. To allow them to pass on costs to the public, companies use a mixture of propaganda and bureaucratic techniques such as lobbying and the formation of alliances with other private and public sector entities.

The key alliance for the nuclear industry is the one with the government. In chapter 4, I explain why government support is critical to nuclear power, describing the many ways in which governments undergird the nuclear enterprise, such as by providing subsidies of different kinds and skewing electricity markets in favor of nuclear energy. This happens both in countries where the electricity business is privatized (e.g., the United States and the United Kingdom) and in countries where the entities operating nuclear reactor are publicly owned (e.g., India and China). A second form of government support is by reproducing pro-nuclear propaganda. Governments often justify investing in nuclear energy by echoing the industry's talking points. They also foster the perception that building nuclear plants connotes higher status, a particularly attractive argument in developing countries.[16]

In chapter 5, I elaborate on the close connections between the production of nuclear energy and weapons. By using examples from different countries that possess nuclear weapons, I show the multiple ways in which the pursuit of one goal can advance the pursuit of the other. The linkage with nuclear weapons also helps explain government support for nuclear energy despite the myriad problems associated with the technology. This should cause concern, because expanding nuclear energy will make eliminating nuclear weapons and avoiding nuclear war much harder.

One of the thrusts of the recent push for nuclear construction to counter climate change has been the claim that new reactor designs will not suffer from the historical problems afflicting the technology. In chapter 6, I explain why new reactor designs will continue to be expensive, risk severe accidents, contribute to nuclear weapon proliferation, and produce radioactive waste. Most of these reactor designs are merely theoretical concepts, existing primarily on PowerPoint slides using computer-generated graphics. It will take decades to commercialize them, even if people were willing to pay the much higher costs involved. I also discuss how prominent billionaires like Bill Gates could profit from investing in nuclear energy, and why they are ideologically attracted to the technology.

In the conclusion, I discuss some features of our electricity system and how it would have to change to accommodate the continuing increase in energy supplied by wind and solar plants. These produce outputs that depend on how fast the wind is blowing and the intensity of sunshine. Managing a grid with such sources of power poses challenges, but I explain why continuing to maintain nuclear plants or fossil fuel plants is inappropriate. Matching the varying outputs of wind turbines and solar panels necessitates flexible responses. But that goes against the economic logic guiding the organizations that operate nuclear and large fossil fuel plants.

I end with the political nature of nuclear power and how it functions best only under a social and economic system oriented toward unrestrained material expansion—the underlying cause of the climate crisis. It is being promoted by powerful elites in governments and businesses who benefit from the current inequitable distribution of resources and power precisely because it comes with the promise, even if it will be ultimately a false promise, that the economic system can continue more or less along the same path while avoiding the dire predictions associated with large-scale climate change. Talking about nuclear power from new reactors serves to delay dealing with the climate crisis. Procrastination might be the thief of time, but it is good business strategy for companies that profit from the current system.

The goal

As I have described above, my book offers a technical, economic, and environmental critique of nuclear energy while locating it within the larger political economy of energy and society. The results of this last exercise add to my fear about the climate crisis. In my view, nuclear power provides insight into the ideologies of those who dominate how today's world operates and how these elites deal with challenges to the current system—namely, by doubling down and misdirecting those concerned about climate change by offering false solutions.

Although my focus is on the specific case of the energy system, there are parallels here for other systems as well. I am following the great African American writer James Baldwin, who argued on the pages of the *New York Times* in 1962 that we must try to "utilize the particular in order to reveal something much larger and heavier than any particular can be."

The particular attributes of nuclear power reveal patterns of action prevalent in the larger economic and political landscape. This book's deeper purpose, then, is to demonstrate through the example of nuclear energy why it is important to understand the climate catastrophe to be as much a crisis of political economy as it is a crisis of ecology resulting from technological choices.

1

Undesirable: Risks to the Environment and People's Health from Nuclear Energy

Ira had just undergone a second operation. She told me that her mother had the same type of cancer, and that recently "the doctors found a 'knot' (vuzol) in my little sister's thyroid as well." Ira, like the other girls, marked the progression of her disease by counting the number of "knots" forming in her throat, chest, and neck. "The doctors tell me how many I have at a given time," Ira said, as if she was engaged in a ritualistic form of anticipation.

Adriana Petryna, *Life Exposed: Biological Citizens after Chernobyl*

Electricity is but the fleeting byproduct from atomic reactors. The actual product is forever deadly radioactive waste.

Kevin Kamps, 2016

Rather than cremating her beloved dog Matsuko, Mizue Kanno chose to bury her pet intact, under a cherry tree in her garden in the town of Namie in Fukushima prefecture. Kanno's hope in doing so was that someday she might find an expert who would exhume and analyze Matsuko's dead body, to establish the cause of death.

Matsuko had died from internal hemorrhaging. That Kanno knew. But when she had asked the vet why her dog had started hemorrhaging, she got no answer. The doctors were not prepared to even do an autopsy.

Kanno suspected that the dog's ill health might be related

to exposure to radiation from the devastating accident at the Fukushima Daiichi nuclear plant that started on March 11, 2011. Namie lies within twenty kilometers of the plant and the town was blanketed with radioactive materials.

At that time, no one knew what they were to do. They had not been warned that their neighborhood nuclear plant could melt down. The mayor of the town received "no instructions from Tokyo" and was not offered "aid or transport."[1] Kanno, her son, and their dog stayed at home till March 15, four days later. When they reached Koriyama, another town within Fukushima prefecture, they were screened for radiation exposure. Kanno's levels were very high. She later remembered that the Geiger counter's needle shot up to 100,000 counts per minute, a measure of the radiation intensity, when it was held next to her jacket and hair. This count was nearly eight times the threshold level used by Fukushima prefecture to decide whether thyroid scanning was to be recommended.

A few years after Matsuko died, Kanno herself was diagnosed with thyroid cancer. The surgery that followed was successful but left her with a scar on her neck. Kanno was by no means alone among local inhabitants in suffering from thyroid cancer. As she pointed out, there was even a shop specializing in "clip-on collars for T-shirts and wide choker-style necklaces" because of the demand for accessories to hide the marks left behind by thyroid surgery.

Thyroid cancer has been the most common and quickly visible health impact of the Fukushima disaster. Children were particularly affected. One study published in the journal *Epidemiology* found an approximately thirty-fold increase in the number of thyroid cancer cases among children and adolescents. Over 300 residents of Fukushima prefecture have been diagnosed with thyroid cancer. Of those victims, at least 266 have undergone surgery to remove their thyroid (thyroidectomy). For the remainder of their days, a paper in the journal *Thyroid* tells us, they will have to ingest thyroid hormones and will experience a poorer quality of life, with possible problems like insomnia and fatigue.

In January 2022, six of these thyroid cancer patients, all between six and sixteen years of age at the time of the accident, filed a suit against Tokyo Electric Power Company (TEPCO) seeking a total of ¥616 million (about $5.4 million) in compensation. Evidently, these thyroid cancers have not been fatal to these patients—at least so far—but their lives have been difficult. They have all undergone surgeries and faced recurrences, and four have lost their thyroid glands entirely. One patient's cancer appears to have metastasized to the lung.

The case is complicated because Fukushima prefecture and the Japanese government deny any connection between the cancers and the accident. Their officials typically attribute the increased number of thyroid cancers to "over diagnosis." The United Nations Scientific Committee on the Effects of Atomic Radiation (UNSCEAR) has come to a similar conclusion.

Such denials have an air of plausibility: cancers take years to develop, and the relationship between being exposed to radiation and developing cancer is stochastic (i.e., it is a matter of chance and not certain). These characteristics make it nearly impossible to trace any specific cancer to radiation exposure and make it easy to sow doubts about any relationship between the two. As I discuss later, however, there is strong scientific evidence for a statistical relationship between radiation exposure and developing cancer, as well as a host of other diseases.

Denial by nuclear and government authorities is also a historically tested strategy. Thyroid cancers also spiked after the 1986 accident at the Chernobyl nuclear plant in the Soviet Union. Then, too, multiple national and international groups denied or dismissed any relationship between these cancers and the accident. But within a couple of decades, the correlation became undeniable. As Dillwyn Williams, a leading researcher on thyroid carcinogenesis, explained in a 2002 article in

Nature Reviews Cancer, the spike in thyroid tumors formed "the largest number of cancers of one type, caused by a single event on one date, ever recorded." Between 1991 and 2015, the total number of registered thyroid cancers in Belarus, Ukraine, and the four most contaminated oblasts in Russia was 19,233, according to an UNSCEAR report from 2018.

Thyroid cancers were just the beginning. Studies have found excess numbers of cancers of other kinds, heart disease, congenital malformations, and so on in the region around Chernobyl. Thus, the idea that a severe nuclear accident—such as the one at Fukushima—will not impact public health is wishful thinking at best. More likely, it amounts to the deliberate introduction of "alternative facts," to use Kellyanne Conway's now famous phrase for falsehoods.

A similar example of the mismatch between reality and wishful thinking—or perhaps even deliberate lying—is at play when nuclear proponents make claims about how nuclear reactors are safe and won't undergo severe accidents in the future. Such claims have been proven false earlier: for example, the head of the International Atomic Energy Agency's safety division assured readers of the *IAEA Bulletin*—only three years before the Chernobyl accident—that Soviet reactors were so safe that "a serious loss of coolant accident is practically impossible."

Despite a history of such false guarantees, nuclear energy advocates assert that future accidents are impossible. This is the first topic I explore in this chapter, taking on many of the claims about the safety of nuclear facilities. I will argue that events that cause widespread radioactive contamination—severe nuclear accidents—can never be ruled out. "Safe nuclear power," in the way that advocates of the technology use the term "safe" (i.e., as not capable of undergoing accidents) is an impossibility, especially as nuclear power expands, especially as climate change results in increasingly common severe weather events like hurricanes. And since radiation, even at very low levels, is harmful to people, the expansion of nuclear energy will necessarily result in increased risk to public health and the environment.

The denial strategy adopted by many nuclear energy advocates is two-pronged: first, deny that accidents are possible, and second, deny that accidents are harmful by rejecting the well-documented links between exposure to radiation and health impacts. So, later in the chapter, I briefly document how the nuclear industry employs strategies reminiscent of the tobacco industry's efforts to delink smoking and cancer documented by historians Naomi Oreskes and Erik Conway in their *Merchants of Doubt*. Or the fossil fuel industry's efforts to deny the reality of climate change outlined by authors like Geoff Dembicki (*The Petroleum Papers*).

Finally, I briefly turn to the two other routine impacts associated with nuclear power plants even if they never suffer any accidents. First, I discuss the challenge of dealing with radioactive wastes that are inevitably produced when reactors operate, explaining why it is impossible to demonstrate that these dangerous substances can be safely managed for the millennia that it will take for all these radioactive materials to decay into stable elements. Then, I discuss another activity that is inescapable as long as reactors operate: the mining of uranium, which has led to contamination of land and water in regions around the world, especially in areas occupied by Indigenous communities.

The Fukushima accident

The Fukushima Daiichi plant comprised six nuclear reactors. On that fateful day in 2011, when a massive 9.0-magnitude earthquake set off a tsunami, three of the Fukushima Daiichi reactors had been shut down for regular maintenance. Only three were operating, and these were shut down as soon as sensors in the plant detected the earthquake. However, the nuclear fuel assemblies within these reactors were still very hot, and the radioactive fission products that had accumulated within these fuel assemblies continued producing heat even after the reactor was shut down. Radioactive elements, by their nature, decay, and each decay releases energy that heats the surrounding materials. There is just no way to turn off that heat. Rob Socolow, a physicist and a colleague of mine at Princeton University at the time of the Fukushima accident, described this unfamiliar phenomenon as "the fire that you can't put out."

This phenomenon was critical to what ensued, because the combination of the earthquake and tsunami had damaged all means of removing this heat. With no outlet, the accumulating heat resulted in the temperature of the fuel assemblies increasing. Soon those assemblies started melting, allowing the radioactive fission products to break out of the outer layer of cladding that surrounded the uranium in the fuel. The cladding was made of an alloy of the element zirconium. As it heated up, the zirconium started undergoing a chemical reaction with the surrounding steam and produced copious amounts of hydrogen. Eventually, that hydrogen gas caught fire and exploded. These explosions at the Fukushima Daiichi reactors were watched by millions on their TV and phone screens at that time.

Over the following days, a complex cocktail of radioactive materials escaped from the damaged reactors. Carried by the wind, these radioactive materials were deposited over much of Japan and elsewhere. The result was widespread contamination of the land and ocean.

A majority of the people from the most contaminated regions near Fukushima—roughly 150,000 in number—had to be evacuated; Mizue Kanno and her family were among those. Yet in the days before they moved out of their homes, and while in transit to more distant areas, they were exposed to high levels of radiation from the radioactive gases in the air and from radioactive dust that had settled on the land and buildings.

In the chaos of those early days, it was impossible to measure radiation levels. There were simply not enough radiation monitors on the ground. As a result, estimates of the doses to which Kanno and others like her were exposed will always remain uncertain and inexact. Coupled with debates about the effects of exposure to radiation, the dose uncertainties allowed for intense political battles over estimates of negative health outcomes like the numbers of cancers.

Nuclear advocates have used this state of affairs to introduce doubt about the health impacts of Fukushima. The playbook originated in the tobacco industry during the period when it attempted to counter the growing understanding of the health impacts of smoking. As Oreskes and Conway explain in *Merchants of Doubt*, the tobacco industry used the uncertainty inherent in all scientific endeavors to undermine what the science had uncovered. The strategy is best captured in a phrase from a 1969 memo from British American Tobacco: "doubt is our product."

These tactics notwithstanding, the sheer visibility of the hydrogen explosions could not be ignored. To all but the most diehard supporters of nuclear energy, Fukushima revealed, yet again, that nuclear power plants can undergo severe accidents. Known for its excellent safety measures to deal with earthquakes and other natural disasters, Japan was justifiably reputed as a high-tech and organized society—very different from the collapsing Soviet regime that confronted the Chernobyl accident. That Japan was struggling to cope with the disaster strongly suggested that other places would fare even more poorly.

Yet today we see nuclear power being touted as a safe and reliable way of generating energy. How can this be reconciled with the record of severe accidents?

Safety by rhetoric

The mismatch between past reality and future claims has been best explained by the British sociologist John Downer. An associate professor in risk and resilience at the University of Bristol, John is quiet and understated, but with a wicked sense of humor. As a PhD student at Cornell University, he studied the safety of airplanes—in particular, the regulation of jetliners.

The Fukushima accident drew his attention to nuclear reactors. Although patently different technologies, airplanes and nuclear reactors possess a unique commonality. As John explained in a September 2021 presentation to the US National Academies of Sciences, Engineering, and Medicine, both nuclear power plants and aircraft need to "do active work in order to remain safe." Airplanes are heavy objects that defy gravity to fly. Nuclear plants must contain the fission process and other dangerous materials while operating under challenging conditions: high temperatures, intense pressure, concentrated energy production, damage to materials from radiation, and so on. These technological goals are not straightforward or easily achieved under all circumstances, and so both airplanes and nuclear reactors are capable of severe accidents.

At the same time, the way we think about accidents in the two arenas is different. As John explained,

> We understand that airplanes very occasionally, but nevertheless routinely, crash, even though any specific crash is always unexpected. We accept such accidents as an inevitable cost of the technology and we anticipate them in our plans and institutions. The same logic does not hold in the nuclear sphere, however ... Public decisions about them [i.e., nuclear facilities] are predicated on an understanding that the chance of them failing catastrophically is so low as to be negligible.[2]

The desire to make people think that the likelihood of accidents is negligible is why we hear the constant invocation of "safe nuclear power." In 2010, for example, President Barack Obama announced that to deal with climate change "we're going to have to build a new generation of safe, clean nuclear power plants" alongside opening up new offshore areas for oil and gas development and investing in "clean coal technologies." The qualifier "safe" is meant to reassure the audience and suggest that their legitimate fears—catastrophic accidents—are irrelevant to decision making.

Downer's insight was that the way policymakers and mainstream sources discuss the potential for future nuclear accidents involves two rhetorical moves. The first is to shift the discussion away from the consequences of accidents and focus only on their likelihood. By turning attention away from scary consequences, one tilts the discussion in favor of nuclear energy.

The second rhetorical move, John explained in a 2014 paper in the journal *Regulation and Governance*, is to postulate that the probabilities of accidents at nuclear plants are objectively calculable. This renders nuclear accidents different from phenomena like terrorism. Most people would realize that one cannot quantify the risk of a major terrorist attack. Nuclear proponents want to quantify accident possibilities because they expect this exercise will result in very small probabilities and lead policymakers to ignore this contingency in their planning.

In this way, the idea that accidents have calculable probabilities allows nuclear advocates to dismiss the danger from reactors. For example, NuScale, a US-based company, announced that a reactor design it is developing is so safe that even if it is struck by a hurricane, there is only a minuscule chance—one in a billion for each year of operation—that it will be damaged. NuScale goes on to calculate that if a reactor is struck by a hurricane, the likelihood of a subsequent release of radioactivity is less than one in a trillion (6.6×10^{-14} for the scientifically inclined). Clearly the aim is to placate any concern.

There are good theoretical reasons to be skeptical about the methodology used to arrive at such figures. In a 2016 paper in the journal *Science and Global Security*, Suvrat Raju—whose research on string theory won him the Nishina Award, given to outstanding Asian physicists—has demonstrated rigorously that the existing empirical record of nuclear accidents means that one can simply rule out such extraordinarily small estimates. Although the methods he used are sophisticated, the essential insight is captured by a poster I saw someone carry in a rally following the Fukushima accident: *They say that the probability of a reactor meltdown is one in a million years / In my lifetime, three meltdowns.* The point is simple. Theoretical predictions of stupefyingly low accident probabilities do not square with the empirical occurrence of severe accidents at nuclear reactors.

Learning from accidents

History tells us that reactors of many different designs are susceptible to severe accidents. Accidents have been initiated by external events, equipment failures, and by workers making mistakes. How operators might act is "intrinsically hard to analyze," as an elite group of safety experts explained in their 1978 Risk Assessment Review Group Report to the US Nuclear Regulatory Commission.

And then there is the problem that assumptions made in safety assessments may not be realized in the real world. "The three-dimensional world doesn't faithfully obey manuals," in the pithy phrase from the historian Gabrielle Hecht. Or as Brian Clough, the English football player and manager, quipped, "We had a good team on paper. Unfortunately, the game was played on grass."

Consider what happened at Japan's Kashiwazaki-Kariwa nuclear plant during the 2007 Chūetsu earthquake. When the ground subsided, various underground electric cables moved downward and created an opening in the reactor's basement wall, which then allowed some radioactive materials to escape into the sea. The failure "was beyond our imagination," a Tokyo Electric Power Company official confessed.

The TEPCO official was not unique in being unable to imagine such contingencies. The underlying problem was identified by sociologist Charles Perrow. After years of focusing on the world of organizations, Perrow turned his attention to nuclear safety when he was requested, as a social scientist, to offer some input into a study of the reactor meltdown at the Three Mile Island nuclear plant in 1979. He traced the origin of the accident not to an individual failure—for example, a mistake made by an operator—but to the structural characteristics of the system. In his book from 1984, Perrow coined the term "normal accident" for such events.[3]

Perrow identified two relevant characteristics, interactive complexity and tight coupling. Their combination makes nuclear reactors and similar technologies prone to catastrophic accidents. It should be intuitively obvious that complex systems can fail more easily—simply because there are more parts that can fail. At nuclear reactors, valves can get stuck, pumps might not circulate water when switched on, and pipes might corrode and break. Perrow's insight was that these failures could combine to create more unmanageable failures due to the capacity of the different parts of the system to interact with each other. There is a greater potential for hidden and unexpected interactions between the different component parts. As a result, the number of pathways leading to a severe accident becomes greater—and beyond the comprehension and predictive abilities of designers. The sheer complexity of nuclear reactors lends itself to the problem of the "unknown unknowns," the phrase made famous by former US Defense Secretary Donald Rumsfeld.

If a potential route for an accident is not foreseen, then the reactor, or any other technological system for that matter, cannot incorporate safety features aimed at protecting against that specific failure route—at least, as designed. And as the 1978 Risk Assessment Review Group Report to the US Nuclear Regulatory Commission pointed out, it is conceptually impossible to list all possible pathways to accidents.

The second characteristic, tight coupling, refers to the lack of buffer or slack between different events that occur in succession. Imagine that you are flying from, say, Boston to Colombo. Odds are that you will not find a direct flight and will have to change flights in, say, London's Heathrow airport. If the gap between the incoming and the outgoing flights is just twenty minutes, then even a small problem could result in a missed flight. For example, you might forget to zip up your backpack. Then, as you are running from one gate to the other, the contents might fall out. You stop to gather everything, put it all into the backpack, and zip it up before going on. Those few minutes might make a big difference to your ability to catch the flight to Colombo. However, if you had scheduled the next flight with a gap of six hours, then your mistake—not zipping up the bag—would not make a big difference. The former itinerary, with just a short gap between the two flights, might be described as a tightly coupled one.

Both characteristics are present in nuclear reactor designs. While reactor designs, of course, differ in many ways, they all end up being complex in their own ways, with parts that interact with each other and the system being tightly coupled.

More safety systems won't eliminate accident risk

The most common approach to reducing the risk of accidents is by adding safety systems, or by introducing redundancy so that there will be a backup in the event that one component fails. But if one starts with Perrow's ideas—in particular, the role of interactive complexity—it becomes clear that many accidents are emergent, not present in any individual component. It may be the case that an additional safety system can compensate for that individual component failing. But because the addition will increase the complexity of the system, new pathways to an accident will emerge. This possibility is best understood through an example first pointed out by the political scientist Scott Sagan in his book *The Limits of Safety*: the case of the 1966 accident that ruined the Fermi fast breeder reactor in the United States.

The accident started with two pieces of zirconium breaking off from what was called the "core catcher" at the base of the reactor. The core catcher was a safety system meant to stop molten fuel from escaping out of the reactor in the event of an accident. Following their detachment, these zirconium pieces blocked channels through which molten sodium, the material that was used to cool the fuel, would flow. Bereft of any means to conduct away the heat produced by fission reactions, those fuel rods melted and contaminated the reactor with radioactive materials. In other words, an additional safety system *caused* the accident.

Adding extra safety systems is no panacea. In his book *Rational Accidents*, John Downer draws on a voluminous literature on safety, specifically aircraft safety, to elaborate on the ways in which redundant systems can fail. Two of those ways are noteworthy and familiar: common cause failures and propagation.

The first, common cause failure, involves the same initiating event disabling multiple options that might have been considered redundant. Imagine you are planning an important visit to a place a thousand kilometers away. You book an airline ticket but figure that you can always drive there if push comes to shove. But then, a day before your trip, there is a major snowstorm. Your flight is canceled, along with dozens of other flights. You also can't drive: snow and sleet have rendered roads into perfect settings for cars and buses to skid, and government officials have asked people to keep off the highways, reserving them for emergency vehicles. You have no choice but to cancel your trip—in other words, your safety systems failed.

To come to a nuclear example: at Fukushima, the earthquake proved to be the common cause. The earthquake took out the external electricity supply to the reactor while also setting off a tsunami. The tsunami, in turn, rendered inoperable the diesel generators that were to provide backup power. In aviation, the case of what happened to British Airways Flight 009 on June 24, 1982, illustrates the capacity of one event to act as a common cause failure. As it was flying over the Indian Ocean, Flight 009 ran into ash from a volcanic eruption, and all four of its engines stopped functioning.

The second problem with redundancy is propagation—wherein failures start cascading from one system to another. When the Fukushima reactors' cooling systems failed, that caused the zirconium cladding to melt, thereby allowing radioactive materials to escape from the fuel assemblies. In the case of proposed small modular reactors (see chapter 6), the presence of multiple reactors at one site means that an accident at one could trigger an accident at an adjoining one.

A final reason why safety systems won't eliminate the risk of accidents is that the external world might act in ways to destroy or deactivate these systems. I write this amid Russia's brutal attack on Ukraine, and its occupation of the Zaporizhzhia nuclear plant. Starting on March 3, 2022, when shelling of one of the buildings in the nuclear complex led to a fire breaking out, there has been international concern about an accident at the plant.

Analysts have come up with at least three scenarios that could plausibly end with radioactive materials escaping into the atmosphere from some facility at the Zaporizhzhia complex and contaminating the surrounding region. The most obvious one is that one of the reactors could be damaged by a rocket or missile. The second scenario involves one of the spent fuel pools—structures filled with water where the irradiated nuclear fuel rods are stored for cooling—being damaged, causing the water to leak out and the fuel rods to burn. The third scenario could unfold if the electricity supplied to the plant is interrupted, perhaps because Ukraine's energy grid collapses, and the plant loses all backup means to cool the reactor—akin to what happened at Fukushima Daiichi. In all these scenarios, systems that are meant to ensure the safety of the fuel would be damaged by external events.

The Ukraine conflict might be the first example of nuclear power facilities being in a war zone. (The only close precedents are Israel's illegal bombing of reactors in Iraq [1981] and Syria [2007].) But the war in Ukraine is unlikely to be the last time nuclear power plants will be attacked. And this does not bode well for their operations. As Michael Sailer from the Öko-Institut in Germany, who was chair of the country's Reactor Safety Commission from 2002 to 2006, pointed out in the 2022 edition of the World Nuclear Industry Status Report (see chapter 2), a nuclear plant needs "a stable environment" to operate safely, including "permanently functioning cooling," which is required even when the reactor is "shut down." No one can ensure that these conditions will obtain during a war.

Another external trigger, ironically, is the problem that nuclear power is meant to address: climate change. An important result of climate change is the increasing frequency of extreme weather events. Many of these affect the availability of water, which is critical to the functioning of nuclear power plants; large quantities of water have to be circulated through nuclear reactors in order to remove the tremendous amounts of heat produced in their radioactive cores. This is why nuclear reactors are almost always located near a large body of water— the ocean or a large lake or river. Therefore, droughts and water shortages, as well as extreme heat leading to a rise in temperatures of water bodies—all of which become more frequent as a result of climate change—can affect the functioning of nuclear reactors.

Other consequences of climate change that affect nuclear plant safety include greater levels of flooding, strong storms and hurricanes, and wildfires. Such events can disable multiple safety systems simultaneously, thereby threatening the safe operation of nuclear plants. Further, in the event of an accident, some of these external conditions—for example, floods or wildfires—would make accessing the site harder, challenging potential responses to the accident. Plant operators, then, might have to shut down nuclear reactors more frequently as a precautionary measure. My former colleague Ali Ahmad showed that in the last decade (2010–2019), the frequency of climate-related nuclear plant outages was already nearly eight times higher than it was in the 1990s. Outages will become only more frequent in the future.

Such shutdowns will have another impact, as each closure reduces the revenue for the organization operating the plant. Andrei Covatariu, Ali Ahmad, and I quantified these losses in the case of Western Europe, and climate-change-related stoppages could result in losses of hundreds of millions of dollars or even over a billion dollars.[4] Such losses will inevitably create pressure to cut costs; should these organizations succumb to such pressure, they will increase the risk of accidents.

Institutions and safety

If nuclear power is so hazardous, what demands might be reasonably made of the organizations operating nuclear reactors and related facilities? And are these organizations likely to meet these demands?

Following the Fukushima disaster, many organizations initiated inquiries into the prior causes of the accident. The most prominent of these inquiries were both called "Independent Investigation Commission," one set up by the National Diet (Japan's parliament) and one organized by the Rebuild Japan

Initiative Foundation (a think tank). Both reports explicitly blamed the institutions in charge of the country's nuclear program.

The commission set up by the National Diet, for example, concluded that the "accident was the result of collusion between the government, the regulators and TEPCO, and the lack of governance by said parties" and went on to express its belief that "the root causes were the organizational and regulatory systems."

The Rebuild Japan Initiative Foundation report found it remarkable that "even in the technologically advanced country of Japan, the government and the plant operator, Tokyo Electric Power Company (TEPCO), were astonishingly unprepared, at almost all levels, for the complex nuclear disaster that started with an earthquake and a tsunami." The report then proceeded to query why preparations proved so inadequate. The most important cause for the failure to prepare was a "belief in the 'absolute safety' of nuclear power," a "myth" propagated by "interest groups seeking to gain broad acceptance for nuclear power."

Although both reports examined only Japan, microscopically examining any country's organizational and regulatory systems would likely find similar lacunae. And like Japan, nuclear advocates globally have propagated safety myths, asserting that accidents are impossible.

An interview with Srikumar Banerjee, the chair of India's Atomic Energy Commission, on the popular television channel NDTV offers a great example, albeit one of many claims about safety made by Indian nuclear authorities:

NDTV: Are Indian reactors safe?
Srikumar Banerjee: One hundred per cent.
NDTV: Can you keep your hand on your heart and say Indian public need not fear Indian reactors?
Srikumar Banerjee: Yes, I am keeping my hand on my heart to

say that Indian public need not fear Indian reactors and if there is any chance of any accident then we will be the first one to say, close it.[5]

The interview took place on March 20, 2011, even as the Fukushima disaster unfolded in confusing and unanticipated ways. It was far too early to even start analyzing the accident or confidently evaluate the risk factors that contributed to it, let alone know whether they would apply to India. Nor are figures like "one hundred percent" ever justifiable, as any scientist should know.

Why would a senior official trained in science make a patently unwarranted assertion? Two years before the Fukushima disaster, safety theorist Nancy Leveson and her team at the Massachusetts Institute of Technology identified one reason in a 2009 paper in the journal *Organization Studies*. The primary mission of organizations operating hazardous technologies, they explained, is typically "something other than safety, such as producing and selling products ... In addition, it is often the case that the non-safety goals are best achieved in ways that are not consistent with designing or operating for lowest risk." Nuclear reactor designers must always trade different priorities: despite lip service, safety is seldom the first priority, and never the only priority.

For most nuclear operators, their mission might be to boost their profits. Or in the case of state-owned entities like India's Atomic Energy Commission, it might be to capture a large fraction of the country's energy sector and achieve concomitant political power. The chair of such an organization would definitely want to avoid closing down reactors, and thus has the incentive to make claims about how safe these reactors are, regardless of the actual risk of accidents.

And then there are non-safety goals set by more powerful actors. In China, for example, the central government's ambitious targets for nuclear energy have meant its state-owned enterprises and regulators have had to rapidly build nuclear plants, even in the face of safety concerns.[6] Turkey's President Tayyip Erdoğan has tried to portray building nuclear plants as proof of his country's stature in the world.[7] For example, his administration explicitly included having two operational nuclear power plants and a third under construction as part of Turkey's Vision 2023, a set of goals laid out for the centennial of the formation of the Turkish Republic. This, of course, meant that these plants would have to be built according to a politically expedient timetable rather than one aimed at reducing risk of accidents. The Chamber of Turkish Engineers and Architects sued the Environment Ministry over its hurried approval of the environmental impact assessment report for the country's first nuclear plant. Of course, as someone who dismissed public concerns following Fukushima with "In that case, let's not bring gas canisters to our homes, let's not install natural gas, let's not stream crude oil through our country," Erdoğan was not particularly concerned about safety.

Political leaders and elites also happily lap up assertions by the nuclear industry about how reactors being built in their country are immune to accidents. Wishful thinking is also seen in other areas. For example, in the realm of economic policy, leaders wanted to believe in imposing austerity measures. The result, as Paul Krugman explained in a June 2013 article in the *New York Review of Books*, was that "papers and economists who told the elite what it wanted to hear were celebrated, despite plenty of evidence that they were wrong; critics were ignored, no matter how often they got it right."

Can regulation ensure that there are no accidents?

One set of institutions has a special responsibility to reduce the risk of accidents: regulatory agencies. Most countries have turned to such bodies to make sense of the risks they are confronted with, and to reassure society about the safety of these technologies. In a 2010 article in the *British Journal of Sociology*, John Downer called such regulators "a twenty-first-century clergy" whose conclusions are typically accepted "at face value with minimal reflection or circumspection."

But it is not just decision makers who look to regulatory bodies. The industries being controlled are keenly interested in how regulators operate, and they do their best to infiltrate or affect these agencies, a phenomenon termed "regulatory capture." One of the problems identified by the independent investigation commission set up by Japan's Diet was the loss of "the necessary independence and transparency in the relationship between the operators and the regulatory authorities of the nuclear industry of Japan," which it clarified was "best described as 'regulatory capture'—a situation that is inconsistent with a safety culture."

Japan is no exception. Writing in the *New York Times*, my former colleague and mentor at Princeton University Frank von Hippel chastised the US Nuclear Regulatory Commission (NRC) as a "textbook example" of regulatory capture. Frank gave multiple examples as evidence, including the case of the Davis-Besse nuclear plant (discussed in chapter 2) and the refusal of the NRC to mandate the safe storage of spent fuel in cooling pools and the installation of filtered vents to capture some of the radioactive materials released during accidents.

The Canadian journalist Matthew McClearn has exposed in the pages of the *Globe and Mail* how the country's nuclear safety commission overlooked dubious data in renewing a nuclear plant's license and allowed the companies it is regulating to continue operating reactors even though components are deteriorating faster than expected. My former colleague Ashwin Kumar and I have elaborated how India's nuclear regulator allowed building of the prototype fast breeder reactor without requiring a containment structure capable of withstanding accidents that are a special challenge for such breeder reactors.[8]

In the United States, the nuclear industry has a powerful lobbying capacity, especially in the form of the Nuclear Energy Institute (NEI). The NEI has weakened the NRC and its capacity to oversee the safety of nuclear reactors by lobbying for changes in rules or by reducing the NRC's budget. The NEI's 2017 End of Year Report proudly announced that it had "worked with the House Appropriations Committee to again reduce the NRC's budget ... by an additional $85 million," going on to explain that this represented a decline of at least "$139 million (close to $800,000 per reactor)" since the 2014 fiscal year.

The nuclear industry also has friends in very high places, especially the US Congress. They have wielded their power over the purse strings of the NRC to weaken the agency. An example was the late senator Pete Domenici. Described as "the most important legislator for all things nuclear" by the former director of Sandia National Laboratories, which designs the non-fissile material components of the US nuclear arsenal, Domenici describes in his 2004 book *A Brighter Tomorrow* how he threatened Shirley Ann Jackson, the NRC's first African American chair, with cutting the "agency's budget by a third" and forcing the NRC to lighten regulations on the nuclear industry. The senator was richly rewarded for his efforts, receiving hundreds of thousands of dollars in campaign contributions, including from "at least three dozen firms on the membership roster of the Nuclear Energy Institute," according to *NBC News*.

Since then, the NRC has been pliant to the industry. The occasional member of the commission trying to forge an independent path is kept in check by other members. When Gregory Jaczko, the NRC chair at the time of the Fukushima accidents, proposed that the NRC postpone licensing of new reactors in the United States till the lessons of the Fukushima accident were internalized, he was rebuffed by other commissioners. His memoir, *Confessions of Rogue Nuclear Regulator*, describes "the bulldozer mentality of the American nuclear power industry and the majority in Congress who supported it."

Another example comes from France, when its nuclear safety regulator, Autorité de sûreté nucléaire (ASN), came in the way of the country's powerful nuclear establishment. When ASN head Pierre-Franck Chevet told journalists that problems at a reactor under construction were "serious, even very serious," retired executives called his action an "abuse of power" and accused him of going against the national interest. This was particularly galling to Chevet, a career civil servant whose motivation to work on nuclear safety dated back to the 1986 Chernobyl accident, which sparked a desire to avoid a similar catastrophe in France.

The inability of regulators to carry out their task should not be surprising. In a 2010 interview with *Talk Nation Radio*, the retired journalism professor Robert Jensen explained that because "the disparity in wealth that is created by the corporate form will inevitably lead to disparities in power" there will never be adequate regulation.

But it is not just corporations. Regardless of their legal status, all sorts of organizations, including state-owned entities, that build or operate nuclear reactors and other facilities wield significant political power. Using this power, such institutions can and do undermine any external bodies overseeing their behavior. Public interest groups can seldom match their capacity.

At a time when capitalists around the world are enjoying success in their decades-long effort to avoid any kind of constraint placed on their activities in the public interest, regulators cannot ensure that nuclear reactors will not have accidents.

Are there other ways of ensuring safety?

Nuclear advocates have other talking points: reactors that depend on passive safety cannot suffer accidents. Passive safety refers to systems that do not need any external input or energy to operate and rely on natural physical laws (e.g., gravity or thermal convection) to remove the heat produced. The argument is that because nothing can interrupt the action of physical laws, these passive systems reliably prevent accidents.

Passive safety is no panacea either. Odd as it might seem, under some circumstances passive systems might be more uncertain in their behavior than the "active" systems they replace. While passive systems might operate, it is not clear whether they will provide the necessary amount of cooling. For example, when compared with electrical pumps that are traditionally used for cooling the fuel during accidents, passive systems might simply not be quick enough at removing the necessary amounts of heat, and the fuel could be damaged.

I certainly believe in the laws of physics—that is expected for someone trained as a physicist. But the same training also tells me that how these laws manifest themselves in various circumstances is not always intuitively clear. Consider, for instance, the smog that forms over cities like Los Angeles. Naïvely, one would think the hot gases and particulates that come out of automobile tailpipes should rise high into the atmosphere, but that does not happen. It is not that the normal law of physics (hot gases rise) stops working, but just that those cities prone to smog typically experience the phenomenon of temperature inversion, usually because these cities are in valleys.

What conditions will prevail during a nuclear accident are not easily predictable in advance. Accidents, almost by definition, are chaotic, occurring due to reasons that engineers fail to consider—an unexpectedly intense earthquake, or flooding caused by a tsunami. As the Fukushima Daiichi's site superintendent, Masao Yoshida, was to say later, "We encountered a situation that we had never imagined."[9] In turn, it is difficult to predict how passive safety mechanisms will work.

During severe accidents, it may not be possible to control what is happening to the nuclear reactor. The situation reminds one of John Steinbeck's characterization of banks in *The Grapes of Wrath*: "The bank is something more than men, I tell you. It's the monster. Men made it, but they can't control it."

Games with numbers

When faced with the historically undeniable fact that nuclear reactors have had severe accidents, advocates for nuclear power deploy a number of related arguments intended to undermine the import of that reality.

Nuclear proponents often argue that the health impacts of reactor accidents are not significant, sometimes going as far as saying that no one died from the Fukushima accident. During one of the side events at the 26th Conference of Parties to the Climate Convention in 2021, the director general of the International Atomic Energy Agency, Rafael Grossi, announced: "No one died from radiation at Fukushima." Recognizing the statement's absurdity, the audience laughed—but at many other venues Grossi might have been applauded.

Grossi and others like him don't explain how they come to this conclusion. One might charitably suppose they mean that no one has died from radiation exposure so far—those in the know would understand the inevitability of the spikes in cancers and other diseases, even if the toll has not yet become evident.

The more likely explanation, though, has to do with the spurious idea that exposure to low levels of radiation does not lead to any negative health outcomes. Despite enormous uncertainty about the radiation dose that specific individuals like Mizue Kanno have picked up, it is nevertheless undeniable that a great many people in Japan have been exposed to abnormal levels of radiation. This is confirmed in reports by international bodies such as the United Nations Scientific Committee on the Effects of Atomic Radiation.

If radiation exposure cannot be questioned, then the next best target for those who want to make accidents like Fukushima seem more benign is to create doubt about whether that exposure will harm people. Therefore, nuclear advocates posit that radiation is harmless as long as it is below a threshold. Some even support a spurious theory called hormesis, according to which exposure to low levels of radiation is actually beneficial. So, people being exposed to radiation does not matter, the thinking goes, as long as the exposure level is not very high.

The idea seems intuitively credible if one compares it with more familiar poisons: small doses are harmless. Some even think a tiny dose of poison might improve health—for example, homeopaths. Radiation, however, is not like these poisons. Scientists and epidemiologists have not found any sound evidence of a threshold below which exposure to radiation is harmless.

Exposure to radiation at any level is harmful, but the relationship between radiation exposure and cancer is statistical, not deterministic. Scientist Jan Beyea has likened radiation exposure to obtaining a negative lottery ticket. Not everyone who gets a ticket wins the lottery, but some do. Likewise, not everyone who is exposed to radiation will get cancer (or some other disease), but some will. At low levels of exposure, the likelihood of cancer is directly proportional to the radiation dose; in mathematical terms, a linear relationship. This is why radiologists speak about a "linear no-threshold" model of radiological harm.

Many scientific bodies around the world have reached this conclusion. For example, based on a "comprehensive review of the biology data," the 2006 report of the US National Research Council's Committee to Assess Health Risks from Exposure to Low Levels of Ionizing Radiation (or BEIR committee, for "biological effects of ionizing radiation") concluded, "The risk would continue in a linear fashion at lower doses without a threshold and that the smallest dose has the potential to cause a small increase in risk to humans."

After reviewing twenty-nine papers that examined "total solid cancer, leukemia, breast cancer, and thyroid cancer, as well as heritable effects and a few nonmalignant conditions," a group of leading epidemiologists concluded in a 2019 article in *Health Physics* that "the preponderance of recent epidemiologic data on solid cancer is supportive of the continued use of the linear no-threshold model for the purposes of radiation protection."

If the relationship between exposure and harm is indeed linear, then anyone in Japan, or other parts of the world, exposed to even low levels of radiation from the Fukushima reactor accidents will have an increased risk of developing cancer. The increase will not be easy to pick out. The cancers from Fukushima-related radiation exposure would occur at the same time as a much larger number of people develop cancers from other causes—for example, smoking or exposure to other toxic chemicals. Nevertheless, some of these cancers will indeed be caused by radiation exposure due to Fukushima.

The BEIR committee has quantified—in a statistical sense—the relationship between radiation exposure and the resulting number of cancers, including how many of them would be fatal. If 100,000 persons were exposed to 0.1 sieverts (the standard unit for effective radiation dose) each—that is, a collective dose of 10,000 person-sieverts—then roughly around 1,000 people will develop solid cancers and around 85 people will develop leukemia; roughly half of the solid cancer patients and three-quarters of the leukemia patients will die.

This is useful in estimating the impact of the Fukushima disaster. According to the 2020-2021 UNSCEAR report, the collective effective radiation dose to the population of Japan from the Fukushima reactor meltdowns over just the first ten years after the accident is 32,000 person-sieverts. Multiplying this collective dose with the BEIR committee's estimates of cancer risk per unit of radiation exposure would result in roughly 1,800 fatal cancers. Despite significant uncertainties, the number is clearly a lot more than zero: many people will die as a result of cancer induced by the Fukushima disaster, if they haven't already.

The numbers in the case of the Chernobyl disaster are much larger. This is because the collective radiation dose to the entire world from Chernobyl has been estimated in a 1993 UNSCEAR report as 600,000 person-sieverts. When combined with the BEIR committee's cancer mortality estimates, this dose estimate will lead to roughly 34,000 deaths. The figure is necessarily uncertain, because of challenges in measuring and translating epidemiological predictions across different populations. But it is far more than zero, or thirty-one, which was the official figure in the Soviet Union for some years.

These deaths will not be evenly distributed. Children, especially infants, exposed to radiation are far more susceptible to cancers than adults are. Men are about 40 percent more likely to develop leukemia and die from these leukemias, while women are over 60 percent more likely to develop solid cancers and 50 percent more likely to die from these solid cancers. These disproportionate impacts have long been discounted by those in power, as Cynthia Folkers explained in a 2021 paper in the *Journal of the History of Biology*. We have decades of activism to thank for an increased recognition of these differences.

Cancer is not the only health impact resulting from exposure to radiation. Epidemiological studies have uncovered evidence linking increased levels of cardiovascular and cerebrovascular diseases and instances of congenital malformations to radiation exposure. One review paper has even suggested that cancer might account for only half of the fatalities related to radiation exposure. But even without getting into the less studied impacts of radiation exposure, just the number of expected cancer cases and resulting deaths in the areas near Fukushima is high.

The debate is deeply political, because the only bodies with the necessary resources to carry out or fund the time-consuming and labor-intensive work needed to produce radiation dose estimates are national or international agencies and institutions connected to nuclear energy. Especially in countries pursuing nuclear power, these agencies have an interest in underestimating doses and impacts. The same is true of international agencies like the International Atomic Energy Agency.

In her 2019 book *Manual for Survival: A Chernobyl Guide to the Future*, historian Kate Brown has elaborated how some of these agencies and scientific administrators used an "arsenal of tactics" in the aftermath of Chernobyl to make unwanted health reports "go away," using a playbook that included classifying data, limiting questions, stonewalling investigations, blocking funding for research, sponsoring rival studies, relating dangers to "natural" risks, and drawing up study protocols designed to find nothing but catastrophic effects. Given this reality, the above estimates of the likely numbers of deaths from the Chernobyl and Fukushima disasters are most likely underestimates.

I should mention one further twist of the "no one died at Fukushima" argument. Nuclear advocates typically dismiss the number deaths from nuclear accidents by comparing it to a similar number of deaths caused by fossil fuel use. In some cases, they go even further and claim that nuclear power saves lives. Perhaps the most publicized version of this comparison came from the well-known climate scientist James Hansen and his collaborator Pushker Kharecha, who estimated in a 2013 paper in *Environmental Science and Technology* that the use of nuclear power around the world "has prevented an average of 1.84 million air pollution–related deaths" and "could additionally prevent an average of 420,000–7.04 million deaths ... by midcentury."

Hansen and Kharecha make a number of questionable technical assumptions in their calculations. But those questions pale in comparison to their use of a classic logical fallacy: a false dichotomy.[10] The calculation is predicated on assuming nuclear power is the only way to displace fossil fuels. Whatever the case might have been in the past, today the most cost-effective and quickest way of reducing air pollution is to replace coal and natural gas with renewable sources of power.

I have made my argument in terms of the number of deaths only because this is the metric most often chosen by nuclear advocates. Even without considering deaths, accidents like the ones at Fukushima and Chernobyl have majorly impacted local people and communities displaced by the high levels of radioactive contamination. The economic consequences of these accidents were also grave and cannot be denied.

Talking about impacts in terms of large numbers does not really do justice to these individuals, their families, and their friends who have to deal with their deaths and illnesses, or indeed to the doctors and medical professionals who may have tried their hardest to save them. Statistics don't bleed, as Arthur Koestler famously observed in 1944 when writing amid the ongoing holocaust in Germany. I agree completely, but my focus is different, and I can say only that there are already many excellent published accounts of these accidents focused on the affected communities and individuals.

Disrupting communities and ecosystems

Many factors conspired to ameliorate the impact of the 2011 Fukushima accident on people's health. The most important stroke of luck occurred at the pool holding the spent or irradiated fuel from unit 4 of the Fukushima Daiichi nuclear power plant. Water in the spent fuel pool started evaporating. Had the process continued, the exposed spent fuel would have caught fire, leading to the release of much larger amounts of radioactive materials than were actually released by the accident. This was part of the worst-case scenario laid out to then prime minister Naoto Kan, leading to the possible evacuation of 50 million residents in the Tokyo metropolitan area. Subsequently, in his book recounting his experience of leading Japan at the time of the Fukushima disaster, Kan wrote that the possibility "sent a chill down [his] spine." Fortunately, evacuation proved unnecessary due to a "fortuitous" occurrence no one could have predicted: water leaked into the spent fuel pool from the reactor well, allowing the evaporating water to be replaced.

The role of luck is not unique to Fukushima, and many nuclear power plants and other facilities have had close calls.[11] We cannot always count on luck, however. Neither does luck change the unsafe character of nuclear power. As Peter Bradford, a former member of the US Nuclear Regulatory Commission, once wrote in an email to me: "the fact that 99% of drunk drivers get home safely doesn't prove that the activity is 'clean, safe and reliable.'"

Another element of luck at Fukushima was the direction of the wind. During the period when the reactors were actively expelling radioactive materials into the atmosphere, the wind largely headed out into the ocean. As a result, much of the radioactive fallout did not affect areas inhabited by people—rather, marine life was more exposed.

One deliberate choice helped too—requiring or advising people to evacuate their homes and the prefecture. This was not done as quickly as it might have been, but this was a case of better late than never, because as the UNSCEAR report records, evacuation averted significant radiation exposures.

Estimates of how many people were evacuated varies from 146,520 residents (according to the Fukushima Nuclear Accident Independent Investigation Commission), to 164,865 people (as of May 2012, according to the Citizens Nuclear Information Center). The idea of leaving the place where they had lived all their lives must have been wrenching for these people. But the high levels of radioactive materials, especially cesium-137, contaminating the land allowed little choice: had they remained where they were, many more would have suffered from and succumbed to cancers and other diseases.

Of course, evacuating hundreds of thousands of people is not an option to be considered lightly. It is particularly difficult, if not impossible, to evacuate certain groups, like hospitalized patients. And any evacuation of this scale, especially under the larger circumstances of devastation due to the earthquake and tsunami, would have resulted, and did result, in many mishaps.

None of this would have justified not evacuating the inhabitants of the areas near Fukushima. It just reinforces the need to prepare and practice for such contingencies. Too often, emergency plans take on the character of what sociologist Lee Clarke aptly termed "fantasy documents" in his 1999 book *Mission Improbable.* These unrealistic plans, Clarke explains, are written and adopted primarily to inspire public confidence in the operating organizations.

Evacuation may mean a one-way move. More than a decade after the 2011 disaster, the Fukushima prefecture retains radioactive hotspots, despite billions of dollars spent on decontaminating the area. Farmers who moved back to the area, anthropologist Maxime Polleri found during his field work, were forced to learn how "to live with contamination," including by carrying radiation monitoring devices.

The Todokedori Radiation Testing Center is in Minamisōma, a city with an estimated population of about 50,000. There, in a room filled with specialized equipment, volunteers test summer crops and other foods that people are hoping to eat. Thomas Bass, a journalism professor who visited the laboratory, wondered, "Is this what our future looks like? A day-care centre full of radiation maps and equipment for monitoring our contaminated Earth?"

Regardless of what the Japanese government or the nuclear establishment says about how successful recovery has been, the lives of the people who lived near Fukushima are forever disrupted. Likewise, in the areas surrounding Chernobyl, the 1986 accident has left thousands of square kilometers uninhabitable because the land continues to be contaminated by the radioactive element cesium-137, which emits penetrating gamma rays as it decays with a half-life of thirty years. People living in the area will absorb these gamma rays as they go about their daily lives. This is what happened to Russian soldiers who occupied this area and dug trenches in 2022.

An even larger area, over 10,000 square kilometers initially, was contaminated with somewhat lower levels of cesium-137. While authorities deemed the radiation levels in these areas were not high enough to evacuate people, people in these so-called radiation-control zones had to change what they ate (avoiding locally grown food, for example) and their behaviors (reducing the time they spent outside, for example).

People are, of course, not the only victims of these accidents. At both Chernobyl and Fukushima, scientists have observed a variety of impacts on the flora and fauna. The two scientists who have been most involved in documenting changes to flora and fauna in both areas are ecologists Anders Møller and Timothy Mousseau. Along with their teams of researchers, the two scientists have documented several abnormalities, including mutations, sterility, white (albino) spots, and population changes.

One example might suffice to illustrate how our nonhuman fellow beings are suffering from nuclear catastrophes. Mousseau explained that relative to what standard models of bird populations would have predicted, there was approximately only one-third as many birds and only half as many species present in high-contamination areas. One reason for the depletion is that many species of birds suffer from declines in sperm counts, sometimes to the point of vanishing; even the available sperms were less viable and had lower swimming velocities compared with birds from areas away from Chernobyl.

One of the more moving sections of *Chernobyl Prayer*, the history of the disaster penned by Svetlana Alexievich, Nobel Prize winner in literature, dealt with the sorry plight of pets, especially dogs, that were left behind as people were evacuated.

I, for one, felt that the most powerful part of the HBO mini-series *Chernobyl* was the scene in which the dogs left behind had to be shot.

And for those who evaluate things in monetary terms, there is the purely financial cost. These have been much better estimated in the case of the Fukushima accident. Compared with the government of the USSR after the Chernobyl accident, which took place as the Soviet state was rapidly declining, the Japanese government took more steps to deal with the impact of the accident at Fukushima. According to the Japan Center for Economic Research's estimate from 2019, the final costs of cleanup may exceed ¥80 trillion (around $750 billion at 2019 exchange rates). This figure does not include indirect costs—for example, the need for replacement electricity or the decline in revenue from tourism.

It is no exaggeration to describe accidents like the ones at Chernobyl and Fukushima as disasters. People living near these reactors had to flee their homes with what little they could gather in a matter of hours, knowing that they may never be able to come back. Ecosystems and landscapes have been poisoned for decades to come. The ongoing disaster continues to disrupt many lives.

All of this could happen to any nuclear plant. Residents of areas near these plants must live with the prospect of things going out of control on one very bad day, having to evacuate their homes, and never being able to return. This scenario is not likely to occur frequently, but it has happened, and it can certainly happen in the future.

Routine environmental impacts

If nuclear reactors face the risk of severe accidents because events can evolve rapidly, even explosively, another environmental risk associated with nuclear power derives from a very slow set of events: radioactive decays and the release of alpha, beta, and gamma rays. Each uranium or plutonium nucleus that fissions within a reactor gives rise to lighter radioactive elements called fission products. Additional radioactive substances are produced when the reactor's parts absorb the neutrons produced during the fission process. For example, iron and nickel nuclei in the steel components absorb neutrons to become radioactive variants of these elements.

Some of these radioactive materials produced in nuclear reactors will continue to emit radiation for millions of years. Two examples are the fission products iodine-129 and cesium-135—close cousins of iodine-131 and cesium-137, which have been responsible for the largest health impacts from the Chernobyl and Fukushima accidents. The numbers 129, 131, 135, and 137 represent the number of neutrons in the nuclei of these elements, which determines how quickly or slowly the nucleus decays. While iodine-131 has a half-life of eight days, iodine-129 has a half-life of 15.7 million years; cesium-135 has a half-life of 2.3 million years in comparison with the thirty-year half-life of cesium-137. Humanity had never encountered these materials prior to the 1940s, when these were first produced to make the bombs dropped on Hiroshima and Nagasaki.

Exposure to the radiation these substances emit when they decay is harmful to health, as we have seen. Therefore, wastes containing these elements must be stored and isolated from human contact for these unimaginably long stretches of time.

Nuclear advocates reject concerns about radioactive waste. Bill Gates is a good example. In a February 2023 interview with *CNBC*, he dismissed waste as "not a huge problem," because it can be put into deep boreholes underground "where it stays geologically for hundreds of millions of years." Gates was referring to the main technical approach that almost all countries with nuclear power plants are planning to use in the future: to build what is called a geological repository. But the confidence expressed by the founder of Microsoft is misplaced. We don't really know, and we cannot know, if the waste will really stay put for hundreds of millions of years.

Uncertainties and unknowable futures

Nuclear organizations and scientific bodies have theoretically studied geological repositories since the late 1950s. Their studies are the basis for assertions like the World Nuclear Association's "Safe methods for the final disposal of high-level radioactive waste are technically proven."

But the term "proven" is inappropriate when talking about repositories. The basic challenge stems from the long periods of time it would take for these radioactive wastes to decay. Over this period, radionuclides will almost certainly migrate into the biosphere. As my UBC colleague Allison Macfarlane, former chair of the US Nuclear Regulatory Commission, once explained, no "site will ... contain nuclear waste indefinitely."

Engineers and scientists have developed sophisticated mathematical models to predict how long it would take some of these radioactive wastes buried in repositories to start contaminating drinking water and agricultural crops. Folding these results into assumptions about how people far into the future will behave, modelers then calculate radiation doses to our far-removed descendants. Nuclear advocates rely on this procedure to make claims about the safety of repositories. But are these calculations reliable?

Empirically validating the predictions of these calculations is impossible. None of us will be around to verify whether these different moving parts—radioactive substances, geological formations, water flows—will behave according to twentieth- or twenty-first-century mathematical models. As Rod Ewing, a professor of geological sciences at Stanford University, explained, "We will never see whether we were correct or not," and we have no means of obtaining feedback.

In 2001, the Committee on Disposition of High-Level Radioactive Waste Through Geological Isolation, Board on Radioactive Waste Management, convened by the US National Research Council, recognized the persistence of "surprises" and had to admit that "there always will be uncertainties about the long-term performance of the repository system." The committee also warned against presuming that these "can be reduced or eliminated by further research and development."

Conditions at the repository can change considerably over the eons. To start with, the geology itself. Geology is a historical science, not a predictive one. Past geological phenomena cannot tell us what will happen in the very long-term future. While the chosen location might have no volcanoes now, there is no guarantee that one might not erupt tens of thousands of years into the future, bringing all those buried wastes right up to the surface.

The surrounding rock is also affected by the presence of the wastes. As the radioactive wastes decay and generate heat, the rock will become hotter. The rock is already not in its natural state, because it will fracture as gigantic machines bore tunnels into it to emplace the wastes. Thus, its ability to prevent water flows could be compromised. Climate change can also affect the flow of water into the repository. These changes might lead to water corroding the containers inside which the wastes are placed. There is also uncertainty about how microbes might impact the repository over these long periods of time.

Finally, we cannot be certain how people will behave. Could people living in, say, the thirty-ninth century start digging in the vicinity of the repository in search of some mineral commonly used then? Perhaps something akin to vibranium in the Black Panther movies? Despite quixotic research funded by the US Department of Energy and other nuclear organizations, one simply cannot reliably communicate the dangers of buried radioactive waste to people living many millennia in the future.

All told there are many known uncertainties. There will also be "unknown unknowns." Therefore, how nuclear waste will behave in the far future is unknowable.

The experience so far

We do not need to wait for millennia to be concerned. Existing repositories have already demonstrated multiple kinds of failures: design failures, human failures, and institutional failures. Take the example of the Asse repository in Germany. Planners built this repository inside a salt dome, ignoring warnings about flooding raised by local NGOs. And sure enough, there has been influx of brine into the repository since 1988. The result of this design and institutional failure was having to retrieve all the radioactive waste buried there at immense expense.

More recently, in 2014, a drum of transuranic waste stored underground at the Waste Isolation Pilot Plant in the United States exploded, releasing plutonium and americium, which made their way to the surface. The price tag was over $2 billion. The explosion occurred because of a decision to use "kitty litter made out of wheat instead of clay," an error that happened amid organizational pressures on workers to accelerate their performance, creating stress and increased workload. Even the Department of Energy concluded that organizations involved in managing the facility had allowed safety culture "to deteriorate within pockets of the organization." Reinforcing the argument about why it is impossible to be confident about safety, three scholars from Stanford University explained why the accident showed "how difficult it is to predict potential failures of such a disposal system over millennia" in an article in the journal *Nature*.

This history of failures suggests that we should expect many more failures to occur during the decades needed to construct a geological repository and emplace hundreds of tons of highly radioactive waste inside. The millennia over which the waste will remain hazardous offer an even longer time for things to go wrong.

Failed plans to construct repositories

Although the nuclear industry and allied organizations have proposed storing waste from nuclear power plants in geological repositories for decades, not a single one is operational. Some countries, like Finland and Sweden, have chosen sites, but these have not been built and certainly have no wastes stored in them.

National plans to site repositories have faced significant opposition. Scholars Gordon MacKerron and Frans Berkhout coined the catchy term "DADA"—short for "decide, announce, defend, and abandon"—to describe how the process has unfolded in many countries: a decision is made by technical personnel with little or no input from affected communities, followed by a public announcement, followed by a defense of the decision against criticism of different kinds, and ultimately an abandonment of the idea in the face of public protest.

Underlying the resistance is what most people think about nuclear waste and repositories to store them.[12] Substantial majorities of people consider nuclear waste with dread and disapprove of plans to dispose of radioactive wastes near them or, often, far away. Plans to store nuclear waste in an area create such a negative image that it leads people to shun or avoid that area, a phenomenon termed "stigma." As a resident of Ignace, one of the areas considered for Canada's nuclear waste repository, explained: "No matter how safe the project is purported to be ... the very idea of Ignace as a nuclear waste 'dump' will sully its name." Thus, if the siting process is to be carried out democratically, with people being given accurate and comprehensive information, most repository proposals will inevitably be blocked by public opinion.

The public's sentiments are often dismissed by nuclear proponents as ignorance. This is similar to how concerns about potential harm from other pollutants have been traditionally dismissed. Institutions that generate technical assessments of risk are not neutral parties without biases, and their claims are intended to serve their underlying interests.

Proposed alternative solutions to the problem of nuclear waste are similarly tainted. Case in point: the chemical reprocessing of spent fuel, which is favored by many nuclear advocates as the best way to deal with nuclear waste. Originally developed during the Second World War to produce plutonium for the atomic bomb that flattened Nagasaki, reprocessing uses chemical means to separate plutonium and uranium in spent fuel from radioactive fission products. The plutonium thus extracted, these advocates argue, could be used as fuel in specially designed nuclear power plants. Reprocessing, thus, also becomes a way to advocate for building more nuclear reactors.

Proponents of reprocessing even have a great PR term for it— "recycling," a word dear to corporate greenwashers. But the term is misleading, because reprocessing does not in any way allow reusing the vast majority of radioactive substances produced in nuclear reactors. The reason is simple: except for uranium and plutonium, no other element can fission and generate energy. The remaining fission products—for example, iodine-129 and cesium-135—cannot be used as nuclear fuel, and so they still have to be buried in a repository or managed in some other fashion.

Reprocessing also produces waste streams containing radioactive materials in varying concentrations. Streams with low levels of radioactivity tend to be very voluminous, which makes storing them on site very expensive. Therefore, reprocessing plant operators simply release these waste streams into the environment, typically into the ocean, where they can travel widely. Radioactive contamination from reprocessing plants in the United Kingdom and France has been detected as far away as Norway.

The other use for plutonium is, of course, nuclear weapons. Indeed, apart from a handful of scientists and engineers, the vast majority of the world's population learned about plutonium from news of the Fat Man bomb that devastated Nagasaki. In 1974, India exploded its first nuclear weapon using plutonium from a reprocessing plant ostensibly built for expanding nuclear power capacity. Reprocessing, therefore, is also greatly attractive to countries interested in making nuclear weapons. In fact, globally, more plutonium has been produced through "civilian" reprocessing than in facilities marked as being military, as the International Panel on Fissile Materials has documented in its annual reports.

Uranium mining and processing

If burying wastes and hoping they will not resurface anytime soon are the best solutions the nuclear industry has to offer after decades of research, it is ironic that at the other end of the nuclear fuel chain, the industry actively hunts for buried radioactive materials and brings them up to the surface. I am talking of mining for uranium, which occurs in conjunction with other radioactive materials resulting from the decay of uranium nuclei.

The irony lies in the result of such mining: the widespread contamination of land and water. Members of the public have suffered numerous illnesses because they get exposed to different radioactive materials produced alongside uranium ore, although these risks have been overshadowed by the risks to workers involved in these activities.

Mining is just the starting point of the series of activities needed to run a nuclear reactor. Once uranium ore is mined, it must be chemically treated to separate the uranium from other minerals. This process creates large quantities of wastes, usually called mill tailings. Tailings contain a toxic blend of heavy metals and radioactive material: the former category includes molybdenum, arsenic, and vanadium, while the latter category includes thorium-230 and radium-226. The radium-226 decays into radioactive radon gas, which mixes with air and spreads to considerable distances. Dealing with these materials is not easy.

The leftover tailings are usually stored above ground, in artificial ponds of water called tailings dams. This water can percolate into drinking water supplies, introducing radium-226 and other substances, like arsenic. Exposure to these hazardous elements is harmful to health. Arsenic, for example, is a known carcinogen and gives rise to a host of other health problems involving organs like the kidney and the skin. The contaminated water is also absorbed or drunk by flora and fauna.

Contamination can be far more extensive when such dams fail—and dams have repeatedly failed around the world.

Collateral damage

In July 1979, one such failure at Church Rock, New Mexico, resulted in over a thousand tons of contaminated sediment and 370 million liters of contaminated water being spilled into the Puerco River, where it flowed all the way to Navajo County in Arizona. As scholar Valerie Kuletz points out in her book *The Tainted Desert*, the "Navajo people in the surrounding area were unable safely to use their single source of water, nor could they sell or eat the livestock that drank from this water." A mere $525,000 was offered as a collective payment to these victims.

Even without accidents, the Navajo people have suffered incalculable health consequences as a result of uranium mining.

But they are, by no means, the only Indigenous peoples thus affected. Much of the uranium that has been mined around the world has come from areas occupied by Indigenous peoples, including in Australia, Canada, India, and the United States. Proposed uranium mining projects, too, tend to be in areas with large Indigenous populations—for example, in Meghalaya in India or in the area around the Grand Canyon in the United States. Naturally, these communities have long resisted uranium mining and associated activities.

The nuclear industry's plans for disposing of radioactive waste streams also disproportionately target areas largely populated by Indigenous peoples. And, again, there has been resistance, and occasional victories. The now-canceled Yucca Mountain repository was strongly resisted by the Western Shoshone people, on whose lands the site is located. In 2020, the Saugeen Ojibway Nation overwhelmingly rejected Ontario Power Generation's plan for a radioactive waste repository near Lake Huron.

Despite the fightback, the nuclear industry continues to target Indigenous communities for siting nuclear waste repositories. Forty years ago, scholars Ward Churchill and Winona LaDuke coined the term "radioactive colonialism," and highlighted that "American Indians" were selected by this process to be "the first twentieth century national sacrifice peoples." Since then, the term "nuclear colonialism"—meaning a "system of domination through which governments and corporations disproportionately target and devastate indigenous peoples and their lands to maintain the nuclear production process," according to Daniel Endres—has become more commonly used.

Colonialism has also been at play in the more traditional sense in the nuclear sphere. France, for example, obtained exclusive access to Africa's uranium reserves as part of the decolonization arrangements it obtained in the 1950s and 1960s, as historian Gabrielle Hecht has documented in her book *Being Nuclear: Africans and the Global Uranium Trade.*

Despite South Africa adopting apartheid laws, the United States played an important role in building the country's first reactor. In contrast, historian Jacob Hamblin points out, the United States not only refused to help Ghana but helped "crush" the "prospect of an ambitious peaceful nuclear program by an independent African nation not ruled by whites."

Nuclear power, like capitalism in general, requires such sacrificial zones and communities. They are typically far from the public gaze and especially far from where the elite typically live. It is not surprising, then, that such elites might think of nuclear energy as a clean source of power. Leona Morgan, a Diné/Navajo anti-nuclear organizer, has rightly criticized the hypocrisy of nuclear advocates like Bill Gates dismissing the environmental impacts associated with nuclear power. As she pointed out on the show *Democracy Now*, he "does not live near an abandoned uranium mine. He doesn't live near a waste site."

Inevitable impacts

Proponents of nuclear energy have to constantly describe their preferred technology using terms like "clean" and "safe" precisely because it is not. No one says safe solar energy or safe bicycles, because the adjective is superfluous. But as history demonstrates, nuclear power is not safe. It is prone to devastating accidents, and the radioactive waste that inevitably accompanies even its routine operations means that the technology can never be clean.

To shift the discussion away from past accidents, nuclear advocates argue that reactors can be safe. The theoretical possibility of safety itself is highly contentious, but that is moot. What we should be interested in is not whether reactors *can* be safe, but whether they *will* be safe. As I have argued earlier, severe accidents can never be ruled out, due to the hazardous nature of the technology and its proneness to "normal accidents."

The likelihood of accidents also depends on actions of the organizations that operate reactors. For nuclear power to significantly contribute to mitigating climate change, a very large number of reactors would have to be built in countries around the world, including in countries with no operating nuclear plants. Can organizations, across countries and cultures, with multiple priorities, including cost-cutting and profit-making, be expected to follow the demanding practices needed to operate these reactors safely?

The answer has to be negative. Exhibit A for this proposition is the entity I started this chapter with—the Tokyo Electric Power Corporation. Its lack of emphasis on safety has been testified to in extreme detail following the Fukushima crisis. If a well-funded organization in a country renowned for its technological prowess, with significant experience dealing with natural disasters like earthquakes and tsunamis, cannot be trusted to prioritize safety, which ones can?

While all technologies might be capable of accidents, the impact of nuclear accidents is without peer. The radioactive contamination resulting from accidents like those at Fukushima and Chernobyl spreads out over large tracts of land. And the land stays contaminated for a very long time. Such accidents become ongoing disasters, not events that can be relegated to the past tense.

For those people who choose to move back to such lands, or those who never leave, living in this contaminated zone becomes a daily challenge. They live in constant concern that they may be told by their doctors that they have cancer. Unless they are like Mizue Kanno, and already developed the dreaded disease.

In November 2022, when Japan's prime minister, Fumio Kishida, announced yet another push to restart the country's nuclear reactors, Tsuyoshi Suda, an activist, told the *Guardian*:

"For Japan to keep putting its faith in nuclear power plants is like a form of self-destruction." Expanding nuclear energy to mitigate climate change reflects the same tendency but magnified many times over.

Even without accidents, nuclear reactors cause a range of environmental impacts—in particular, the production of a whole range of radioactive wastes. There is simply no risk-free way of dealing with these wastes. They are like the ring in J. R. R. Tolkien's *The Lord of the Rings*. In Tolkien's telling, the ring creates a dilemma where "there are but two courses … to hide the Ring for ever, or to unmake it. But both are beyond our power."[13] Likewise, we can neither unmake radioactive wastes once they are created nor bury them in a manner that we can be absolutely sure that they will never come back out. This hard task is made harder by the continued production of more radioactive waste.

In 2012, as India's government was engaged in a crackdown on protestors resisting the construction and operation of a nuclear power plant in Koodankulam, Australia's prime minister visited New Delhi to discuss, among other things, exporting uranium to India. At that time, the Australian Nuclear Free Alliance sent a solidarity statement addressed to the people of India that pointed out the obvious: "On a good day Australian uranium becomes radioactive waste. On a bad day it becomes fallout [from an accident]." That is, of course, true of all uranium, not just what comes from Australia.

The author of that formulation, Dave Sweeney, later added that on a really bad day, the uranium ends up in a weapon of mass destruction dropped on a country. I discuss that relationship in chapter 5.

2

Infeasible: The Financial and Temporal Costs of Nuclear Energy

It is difficult to fault optimism; it is equally difficult to accept fantasy.

Nikhil Desai, 1984

We've had our share of useless presidents, but at least in the past they knew when to listen and when to back down ... But Macron, he's on another planet.

Michel Doneddu, pensioner, at a
March 2023 protest in Paris

In early February 2022, French President Emmanuel Macron visited the town of Belfort in eastern France. Gearing up for his reelection bid, Macron chose Belfort's General Electric factory to give a major speech on energy policy. France's official government Twitter account posted a photo of the president standing against the backdrop of a giant steam turbine, addressing seated executives and workers in masks. Macron used this location to unveil a series of steps aimed at moving away from fossil fuels and reducing France's carbon emissions.

Although the speech included many proposals, and Macron advocated for "a plural strategy ... to develop both renewable and nuclear energies," media outlets around the world overwhelmingly focused on a program of building nuclear power plants, which, the president promised, would "lead to the commissioning of 25 gigawatts of new nuclear capacity by 2050." *World Nuclear News*'s headline—"Macron Sets Out Plan for French Nuclear Renaissance"—was not surprising for an organization whose raison d'être was promoting nuclear energy. But Reuters chose to title its article "Announcing New Reactors, Macron Bets on Nuclear Power in Carbon-Neutral Push," whereas the *Guardian* announced, "France to Build up to 14 New Nuclear Reactors by 2050, Says Macron." And the *New York Times* account carried the headline "France Announces a Vast Expansion of Nuclear Power."

One had to dig deeper into these articles to see that even Macron, despite his clear fascination with nuclear power, had to admit that France needed "to massively develop renewable energies because it is the only way to meet our immediate electricity needs, since it takes 15 years to build a nuclear reactor." Newspapers evidently didn't find that dose of reality worth highlighting.

One reason might be that media outlets had become accustomed to pronouncements by politicians and government officials about the virtues of nuclear power. The UK's Boris Johnson, for example, used the occasion of his very first address to the House of Commons as prime minister to announce, "It is time for a nuclear renaissance, and I believe passionately that nuclear must be part of our energy mix," and express his conviction that nuclear energy will help the UK meet its carbon-emission reduction targets.

Political officials from the next rung have also sung the praises of nuclear energy. In Canada, for example, Minister of Natural Resources Seamus O'Regan pronounced that there was "no path to net-zero without nuclear power." And down south, Jennifer Granholm, secretary of the US Department of Energy, stepped up to say pretty much the same thing at a meeting of the American Nuclear Society.

The delusional nature of Macron's announcement was best explained by Mycle Schneider, an energy analyst based in Paris. In an email to Geoff Fettus, who was then with the Natural Resource Defense Council, and shared with a group that followed nuclear power news closely, Mycle explained that

Macron had talked about constructing a new design called the EPR2 being deployed by 2035. The EPR2 was still at a "basic design" stage and would take an estimated 20 million hours of work by highly skilled engineers and scientists just to bring it to the "detailed design" stage. But so far only about 1 million hours of work had been put into the design. Being a master politician with an eye to the polls, Macron had mentioned the 1 million hours figure in his speech, omitting the inconvenient fact that another 19 million hours was necessary. In practice, the development of a reactor design to the point when it can be licensed and constructed could take even longer. Based on a leaked October 2021 internal inter-ministerial government summary of the status of the EPR2, Mycle pointed out that if everything went smoothly, the first reactor could start operating in 2039–40, and if less well, 2043.

In a follow up message, Mycle also added, "These are strange days. But life in La-La-Land is not new for the French nuclear establishment." He then pasted a graph presented by Électricité de France (EDF) in 2008 at the London "Investor Day." The graph illustrated what EDF termed the "New Nuclear revival" involving a new 140 gigawatts of new nuclear reactor capacity (roughly 140 new reactors) being built between 2007 and 2020. And Mycle added that when the situation in 2020 was compared with that in 2007, overall capacity was slightly below, and there were around fifteen fewer reactors operating.

Mycle's quick recall of the 2008 announcement wasn't surprising to me. Over the years that I have known him, I have always been struck by the very organized filing system he used to archive news materials.

Mycle was, if nothing else, efficient. When I visited his house in the outskirts of Paris in November 2018, he gave me a twenty-minute lecture on his computer setup: two monitors, a separate keyboard, a chair adjusted to the correct height, and so on. This lecture was not occasioned by my complaining about neck aches or carpal tunnel syndrome. What set off

Mycle was his disbelief at how I worked on my laptop, with no external monitor. Switching from one program or one document to another was inefficient, he explained, and could be avoided if I had two screens and could look at them simultaneously. I left the conversation with the distinct impression that all the time wasted through such inefficient practices would have added up to many years of my life.

Mycle's bent for efficiency helps him immensely in what occupies most of his time these days: documenting the abject performance of the nuclear power industry in great detail. Every year since 2007 (and twice earlier, in 2004 and 1992), Mycle has been producing the World Nuclear Industry Status Report, or the Status Report for short, an extensively referenced document that measures how nuclear energy has fared during the previous year against multiple indicators. With time, the team producing it has grown, and so has the report itself: from 34 pages in 2007 to 549 pages in 2023. I have been contributing sections to the Status Report since 2015—a fulfilling but grueling process, in part because of the extensive fact checking involved.

Besides pages and pages of information, what the Status Reports provide are a way to comprehend broader trends. As Mycle explained in an interview to Amy Goodman of *Democracy Now* in April 2011, "You've got to look at the film. Don't look at the photograph. Look at the film in order to understand what's happening."

What would looking at this metaphorical film help us understand?

Whereas a photograph provides a static image, a film captures temporal changes. In the introduction, I discussed one such change: how much nuclear energy contributes to global electricity production. A good starting year is 1997, when countries joined together in Kyoto and adopted the first commitment under the United Nations Framework Convention on Climate Change to reduce their carbon emissions. Then, nuclear energy contributed approximately 17 percent of the world's electrical energy. In 2022, that fraction had come down to just over 9 percent of worldwide electricity. In contrast, modern renewables—namely, solar, wind, geothermal, and biomass-based energy—have grown in importance, from around 1 percent in 1997 to over 14 percent in 2022.

The roughly 9 percent of electrical energy coming from nuclear reactors is already quite small. But, in fact, nuclear power's contribution is even more limited, because electricity is not the only way people require energy. People move around in cars or buses, heat or cool houses and offices, eat food produced in distant places using fertilizer and pesticides, and purchase or use a variety of products made in factories. Many of these activities do not use electricity but use other energy sources. Fossil fuels like coal, natural gas, and oil constitute the largest share of all such energy use. In contrast, nuclear power contributes only 4 percent. (Modern renewables come out at 7.5 percent.) To put it differently, globally, an average person requires nuclear energy for under *one hour every day*.

One way to understand the decline in nuclear energy's share of electricity in the last quarter century is to look at another temporal trend in the metaphorical film Mycle talked about: the construction of nuclear plants. Globally, the number of nuclear plants connected to the grid annually peaked in 1984 and 1985; during each of those years, thirty-three nuclear reactors became operational. The decline thereafter was sharp, as documented in the various editions of the Status Report. The timing of the decline coincided with the Chernobyl reactor exploding in 1986.

Another temporal trend affecting nuclear energy's share is a tussle between new reactors being commissioned and older reactors being shut down, often because they could not compete economically with alternatives. In the first two decades of this century, 95 reactors were started up around the world while 98 reactors were closed down.[1] Between the startups and the shutdowns, the nuclear fleet has stayed more or less constant since the late 1980s.

These trends have implications for generating energy under a carbon constraint. If nuclear power is to contribute meaningfully to climate change mitigation, the fraction of energy it contributes should be increasing with time, with more and more fossil fuel plants being shut down and replaced with nuclear power plants. That hasn't happened, for reasons that I elaborate. The bottom line: as a source of global energy, nuclear power is, if I may indulge in a pun, not glowing.

How does one reconcile this declining trend of nuclear power with the hyperbolic statements of people like Macron and Johnson? Such governments, and private companies from those countries, have invested an enormous amount of money in propelling nuclear development. The enormity of these investments was more than matched by another enormous quantity.

The cost of construction

In the county of Somerset in southwest England, Électricité de France is building a new nuclear power plant. The plant features two European pressurized reactors (EPRs), each of which is designed to generate 1,720 megawatts of electrical power. The two reactors together are to feed 3,260 megawatts into the electricity grid, if all goes according to plan. The remaining 180 megawatts will meet various operational needs within the plant itself: to run cooling pumps, for example.

Construction of Hinkley Point C can be described only in superlatives. About 6 million cubic meters of soil and rock have been excavated for one purpose or the other. EDF announced in 2019 that it had poured 9,000 cubic meters of concrete, reinforced by 5,000 tons of steel, into a large hole in the ground that it had excavated previously. By the time the two reactors are ready, at least 200,000 tons of steel will have been used on that site. Others have claimed that the project might require up to a million tons of steel. Even the electrical power consumed within the plant is nearly as much as a small country; Eritrea's power plants, for example, can together generate only 200 megawatts.

The British media loved these superlatives about Hinkley Point. In 2021, for example, the BBC Two TV channel broadcast a documentary called *Building Britain's Biggest Nuclear Power Station* that measured the project "in swimming pools and football pitches"; extolled "a tunnelling machine so enormous it requires a police cavalcade"; praised the "largest continuous cement pour in the UK"; and informed viewers "that Hinkley's canteens consume 316 tons of baked beans a year."

What the media's focus on such trivia occludes is the cost of the project, another feature that demands superlatives. As of February 2023, construction alone is estimated to cost almost £33 billion (roughly $40 billion). All that steel and those large tunneling machines are expensive! But this figure is by no means fixed and will likely become larger with time. Indeed, it has increased in steps over the last decade: £16 billion in 2013, to £19.6 billion in 2017, to between £21.5 billion and £22.5 billion in 2019. Whatever the final cost, Hinkley Point C will doubtlessly be one of the most expensive construction projects of all time.

There are other nuclear plants that are competing strongly for the same dubious status. Close on Hinkley Point's heels is the Vogtle project in the state of Georgia in the United States, with a total cost estimate of close to $35 billion. That might seem a little lower than Hinkley Point, but not if we account for the fact that each AP1000 reactor is designed to generate only around 70 percent of the electricity that each EPR unit is designed to generate. Like Hinkley Point, the plant's cost has risen in steps, from the $14 billion that was promised when construction was approved.

And then there is the single EPR being constructed in Flamanville in France that is running at €13.2 billion (around $15 billion). This is also in the same ballpark as the other two when the different power outputs of these reactors are taken into account. Flamanville's distinction, though, is that this cost estimate is more than four times what was forecast when construction started. Further, as detailed in the 2020 Status Report, there are other costs, including for financing the project, which could add up to another €6.7 billion.

These projects are not the only instances of expensive nuclear power plants where costs ballooned.[2] Russia's Leningrad-2 plant went up from ₽133 billion to ₽244 billion. India's Koodankulam-1 and -2 reactors, imported from Russia, rose from ₹131.71 billion in 2010 to ₹224.62 billion by 2015, and its prototype fast breeder reactor has gone up from ₹34.9 billion to, currently, ₹68.4 billion.

Initial cost estimates have, in all these cases, proven to be absurdly low. Like sand at the bottom of an hourglass, the cost estimate for nuclear plants invariably keeps accumulating as time goes by.

Reactor construction times also exceed initial estimates routinely. Below I list a few examples from the 2021 Status Report of reactors that were started up between 2018 and 2020. In China, the country that is sometimes held out to be the great hope for nuclear power, eleven reactors become operational during this period. Of these only two—Tianwan-4 and Tianwan-5—met the expected construction schedule. Taishan-1 and -2 reactors were to have taken 4.1 and 4.5 years but took 8.7 and 9.2 years respectively. Likewise, Sanmen-1 and -2 reactors went from an expected 4.5 and 4.7 years to 9.2 and 8.7 years respectively. In South Korea, an erstwhile hope for nuclear power, the Shin-Kori-4 reactor went from a projected 5 years to 9.6 years. In Russia, the country that has led the race in signing reactor export contracts, the twin units at Leningrad-2 were to be completed in 5 years but took 9.4 and 10.5 years.

These details might seem boring and repetitive, but they corroborate the persistent observation that nuclear power plants are seldom built on schedule or under budget. One study by Benjamin Sovacool, Alex Gilbert, and Daniel Nugent examined 180 nuclear projects and found that a mere 5 met anticipated cost and time targets. The remaining 175 took, on average, 64 percent more time than projected, and had final costs that exceeded the initial budget, again on average, by 117 percent. That study examined other electricity infrastructure projects as well; these, too, had cost overruns, but the authors observed that nuclear reactors "stand apart at the top ... for both mean cost escalation and frequency."

Because we are discussing many countries that have built nuclear plants for decades, these cost and time escalations cannot be because of ignorance or inexperience. The only explanations that make sense are all more troubling. A recurrent problem is the underestimation of costs and construction times by project proponents, both when advocating for investment in these projects and during construction. Often, the correct descriptor is not "underestimation" but "deliberate misrepresentation."

Bent Flyvbjerg, who specializes in studying large projects of all kinds, explained in the pages of *Harvard Design Magazine* in 2005 that the ones that receive investment and approval are ones where the "proponents best succeed in designing—deliberately or not—a fantasy world of underestimated costs, overestimated revenues, overvalued local development effects, and underestimated environmental impacts." Flyvbjerg was not discussing nuclear projects specifically, but that diagnosis is spot on when it comes to nuclear projects.

Advocates for expanding nuclear power to mitigate climate change are trapped in such fantasies, unless they are propagating these fantasies themselves.

Operational costs and shutdowns

Building nuclear plants is only the first economic challenge. There is a second economic problem for nuclear power: the high operating costs of plants. The latter includes the costs involved in paying workers, buying uranium, fixing failed equipment, and so on. Together, these set the minimum price the owner of the plant needs to be paid for electricity. Otherwise, the company would operate at a loss.

This is particularly challenging in states and countries that have opened up their electricity sector to market competition. Multiple terms are used for this process, each with its own ideological underpinning: deregulation, liberalization, privatization, and so on. But the basic idea is to treat electricity as one would refrigerators or bananas—that is, as a commodity to be bought and sold on a marketplace. In such markets, the price for the electricity is set by all the plants that can supply this commodity, with their differing production costs.

The net result of introducing competition into the electricity markets has been what George Orwell observed when he reviewed what would become a libertarian bible, Friedrich Hayek's *The Road to Serfdom*: "The trouble with competitions is that somebody wins them." In the electricity sector, nuclear power lost the competition.

This is why, over the past decade, a number of old reactors have been retired, even though they were still licensed to operate for many more years. More generally, the reactor fleet is declining in many countries, as even the International Atomic Energy Agency's annual publication series, *Nuclear Power Reactors in the World*, testifies. In the United States, there were 104 nuclear reactors in operation at the end of 2010. A decade later, at the end of 2020, there were 94. The United Kingdom came down from 19 to 15 reactors; Sweden from 10 to 6 reactors. More reactors are due to be shut down. In all these countries, the main reason for these shutdowns was poor economics. The European utility company E.ON justified its decision to shut down two of its reactors in Sweden by emphasizing the absence of any "prospects of generating financial profitability either in the short or the long term."

If this is the outlook for nuclear plants whose costs have been paid off, it is not surprising that new reactors are simply not competitive in the electric marketplace.

The nuclear industry has an excuse for these shutdowns. Nuclear energy is not valued properly, according to the industry lobbying group Nuclear Energy Institute. Should its benefits be adequately appreciated, electricity companies will pay a lot more for it. Or so goes the refrain.

To those with an interest in history, this narrative should sound familiar. It was the tobacco industry's tune as it faced increased scrutiny from the medical establishment. The 1967 annual report of Philip Morris, for example, proclaimed: "Unfortunately the positive benefits of smoking which are so widely acknowledged are largely ignored by many reports linking cigarettes and health, and little attention is paid to the scientific reports which are favorable to smoking."

Like the supposed benefits of tobacco, nuclear energy's advantages, such as they are, have been proclaimed so loudly by entities with deep pockets that they are in little danger of being undervalued. That excuse for nuclear power can be dismissed outright, just as claims made by cigarette manufacturers during an earlier era were. The economic problems of nuclear power, instead, have to do with the material reasons already described.

Nuclear power's economic problems will worsen as more renewable energy comes on the grid. Solar and wind power plants have low operational costs, which means that it is economically sensible to use their output as and when they are generating power. But their outputs are variable. I will discuss this challenge briefly in the conclusion. For now, all I will say is that the ideal way to compensate for the variability of solar and wind power is a complementary source of power that can also vary quickly.[3]

Though possible, varying the output from nuclear power plants is challenging for a number of technical reasons. More importantly, doing so would decrease their economic competitiveness. If nuclear power plants have to act as complement to solar and wind energy then each plant would sell fewer units of electrical energy. That would increase the cost of generating power because the high fixed expenses at nuclear power plants have to be recouped over fewer units of energy sold. Operating reactors in this manner would also make revenue streams uncertain since how much nuclear power is generated will depend on the ebbs and flows of sunshine and the wind.

The US state of Illinois provides an illustration of this problem. Illinois is among the states most dependent on nuclear energy for its electricity. Dominating its supply is Exelon Corporation, which operates eleven reactors within that state. Speaking at the 2021 World Nuclear Association's annual symposium—in other words, among nuclear proponents—an Exelon official explained the corporate giant's problem was that wind power "coming in from the Dakotas and elsewhere" can "depress the market prices, particularly in the evening whenever the wind is high and the load is low." This competition with wind energy meant losses for Exelon. Exelon threatened to close its nuclear plants, but simultaneously lobbied for subsidies. The lobbying paid off and the state of Illinois chose to subsidize Exelon's nuclear fleet. Naturally, these subsidies are paid for by customers.

It is this difficult economic context that leads nuclear plant owners to desperately seek new sources of revenue, for example, through the bizarre alliance with Bitcoin-mining firms. The relationship is a result of the two enterprises coming together to hope that each other's woes will be the answer to their problems. For cryptocurrency miners, the problem is the extremely energy-intensive nature of their enterprise, which is pumping out more carbon dioxide than some countries. Even if the miners are not genuinely concerned about the environmental impact of their activities, the increasing number of articles in the media that highlighted the emissions associated with cryptocurrency must be embarrassing. For nuclear power plant owners, the problem is selling their expensive power on competitive electricity markets. This has resulted in a spate of announcements about cryptocurrency firms entering into agreements with current or prospective nuclear plant owners to buy electricity from them and claiming environmental brownie points.

There is an obvious problem with this marriage: expensive power won't help Bitcoin manufacturers, and a niche and unstable market won't help nuclear reactor owners. But it is a sign of the times that people think that nuclear power can be used to greenwash an enterprise that is environmentally wasteful and has no social purpose, utilizing huge amounts of energy to do pointless computations. (Unless, like Ted Cruz, the former presidential hopeful from the Republican Party, one believes that Bitcoin will be the solution to strained electricity grids.)

Economically, then, there is no case for nuclear power. It has been, and will continue to be, an expensive way of generating electricity.

Comparing costs

How do these costs measure up with those of other sources of electricity—say, solar energy? This comparison is not so straightforward. One of my local newspapers, the *Vancouver Sun*, told its readers a few years ago that "a rooftop array of 20 photovoltaic panels with a capacity of five kilowatts will cost around CAD 15,000 installed." While that is infinitesimal in comparison to the earlier mentioned cost figures for nuclear reactors, that is an unfair analogy. The solar photovoltaic system costing C$15,000 can generate at most five kilowatts, which is also a tiny fraction of the power that typical nuclear reactors produce.

To compare costs for such disparate power systems, energy analysts often compare the cost per unit of generating capacity (in, say, kilowatts). Thus, the cost of solar power in the above example is C$3,000 per kilowatt.

The two Hinkley Point C reactors can provide up to 3,260 megawatts of electricity. If these were to end up costing £33 billion, then that would translate to C$17,209 per kilowatt (using a conversion rate of 1 British pound to 1.7 Canadian dollars), or nearly six times the corresponding figure for a solar panel on a roof in British Columbia.

Even after that adjustment, this is not quite an apples-to-apples comparison due to at least three main differences between solar and nuclear power.

The first is that the solar panel on someone's house feeds the electricity directly to the household. The nuclear plant requires hundreds of miles of transmission lines, transformers, and distribution systems to supply an individual house or office or factory with electricity.

One way to sidestep this difference is by comparing nuclear power with a larger solar plant that involves lots of panels laid out on a field—what is often called utility-scale solar, because it is usually electricity (utility) companies that build such installations. Putting up solar panels on a roof is more complicated and expensive compared with doing so in fields, and thus the cost difference between nuclear power and utility-scale solar energy is greater.

The financial firm Lazard has been comparing the costs of different energy technologies for over a decade now. In its annual report from October 2021, Lazard calculated that the average construction costs of a utility-scale solar photovoltaic plant in the United States—one of the largest renewable energy markets in the world—was $875 per kilowatt of generation capacity. (For comparison, the cost of a residential rooftop photovoltaic system in the United States was about $2,600 per kilowatt.) These estimates are averages over many different projects and thus smooth over the peculiarities of individual locations, differential labor costs, and geographical variations. Lazard estimated that a nuclear plant costs around $10,300 per kilowatt—or nearly twelve times the corresponding cost for utility-scale solar photovoltaic plants.

This comparison would have been different some years back. The cost of solar photovoltaics was $1,750 per kilowatt in 2013, or roughly twice what it was in 2021. The nuclear trend was the opposite: building a *new* nuclear reactor *rose* from nearly $6,800 per kilowatt in 2013 to $10,300 per kilowatt in 2021. For reasons discussed later in this chapter, it is very unlikely that nuclear plants will become cheaper in the future. Solar costs continue to decline and there is every reason to expect this trend to continue.

The second, and perhaps the more obvious, difference between solar and nuclear power is that a photovoltaic panel will generate electricity only when the sun is shining. In rainy British Columbia, Natural Resources Canada informs us, each kilowatt of installed solar capacity will generate only around 1,000 kilowatt-hours of electrical energy in a year. By comparison, if and when Hinkley Point C starts operating, it might generate around 7,800 kilowatt-hours per year for each kilowatt of capacity, unless it is operated with a varying output.

Again, this is a distinction that energy economists have long dealt with, and they do so by comparing the cost per unit of electrical energy generated rather than the cost per unit of capacity: dollars per kilowatt-hour rather than dollars per kilowatt. Such a comparison would reward the nuclear plant for generating more electrical energy for the same installed capacity as compared with a solar plant.

Renewables like solar energy produce a variable stream of electricity. But this variability does not mean that they cannot be the basis of a grid that has to provide electricity at all times of the day and the year. How this can be dealt with, and why this rhetorical castigation of renewables is misplaced, is discussed in more detail in the conclusion.

There is a third difference that should be addressed to improve the comparison. The timings of the cash flows, to use project finance terminology, differ in the two cases. In the case of solar plants, the costs are dominated by the initial construction, and this is done over a relatively short period—a year or two, typically. Once set up, a solar plant costs little to operate and maintain. That mainly involves cleaning panels (someone with squeegees to wipe away dirt) and replacing pieces of equipment that fail or wear out.

The costs associated with a nuclear plant are expended over an extended period. The first set of expenses starts well before there is any electricity to be sold and involves the cost of the work needed to obtain the requisite permits. The second set includes the expenses discussed in great detail earlier, those associated with constructing the reactor. The third set of cash flows occurs during the period—many decades in the case of well-functioning plants—that the nuclear reactor is generating electricity, or even when it is shut down for repairs or refueling. These operating costs include buying fuel and employing highly trained reactor operators and other workers, as well as security guards to protect against attacks and safeguards inspectors that ensure that no plutonium or enriched uranium is diverted. Finally, there are the expenses to be incurred after the reactor stops generating electricity and revenue: the costs of decommissioning, which involves defueling the reactor, deconstructing it, and then dismantling it; and the cost of waste management and disposal.

The process that is commonly used to account for this variation in the time distribution of costs is called discounting, a practice developed in the mid-nineteenth century by a group of German foresters, and then subsequently rediscovered by the economist Irving Fisher in 1907. Discounting allows future costs to be treated as though they are spent in the present. Through such discounting, energy economists calculate what they call the levelized cost of energy.

In 2021, Lazard estimated that on average the levelized cost of utility-scale solar energy is $34 per megawatt-hour; the corresponding average figure for newly constructed nuclear plants is about $168 per megawatt-hour, nearly five times as high. Wind power came out at $38 per megawatt-hour.

Comparisons in other countries come to similar conclusions. In the United Kingdom, the Department for Business, Energy and Industrial Strategy (BEIS) regularly publishes estimates for levelized costs for different generation technologies. In its 2020 report, BEIS estimated that a large-scale solar project targeted to become operational in 2025 in the UK would produce electricity with a levelized cost of £44 per megawatt-hour. In comparison, its estimate for nuclear power is roughly £102 per megawatt-hour.[4]

The BEIS and Lazard estimates are quite different, especially for nuclear energy. This has to do with differences in

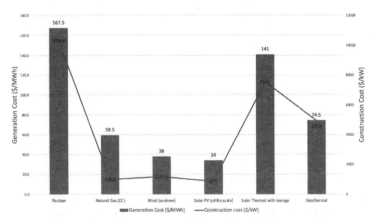

Figure 1: Cost of constructing different kinds of power plants and generating electricity at these. Source: Figure plotted using numbers from "Lazard's Levelized Cost of Energy— Version 15.0." New York: Lazard, October 2021.

assumptions and in financial circumstances in the two countries. But the basic point comes through in both of these, as well as estimates by other groups: nuclear energy is *far more* expensive to generate than solar power. The cost differences between nuclear and renewables are so great that even after allowing for additional costs associated with managing the variable output from solar and wind power fleets, the latter are still cheaper.

Why do these cost differences matter? One constraint on dealing with climate change is the economic impact of doing so. Because nuclear reactors are more expensive, a country investing in nuclear power would be reducing its carbon emissions more slowly compared with a country investing a similar amount of money in renewables.

Several academic studies bear out this piece of common sense, including one econometric analysis of data from thirty countries published in 2018 in *Renewable and Sustainable Energy Reviews*, and another paper from *Nature Energy* in 2020 that examined data from "123 countries over 25 years." The second study went further and addressed an objection that some raise: Why not both nuclear and renewables? By looking at national investments in these two technologies, it found that the more a country invested in nuclear power, the less it tended to invest in renewables, and vice versa. One crowds out the other, and there is not enough room for both. We must make a choice.

In the words of Peter Bradford, a former member of the US Nuclear Regulatory Commission,

> Those who assert that the problem of climate change is so urgent that we have to do everything … overlook the fact that we can never afford to do everything. The urgency of world hunger doesn't compel us to fight it with caviar, no matter how nourishing fish eggs might be. Spending large sums on elegant solutions (especially those with side effects) that provide little relief will diminish what we can spend on more promising approaches.

Investing in nuclear power only makes things worse for climate mitigation.

It takes a lot of time

Nuclear reactors are not just expensive. They take long to construct. Had it been possible to build them quickly, perhaps it would have been possible to increase the amount of nuclear power in the system even as older or more problematic nuclear plants were shut down.

To appreciate the time scale, one can do a thought experiment. Imagine that the government of British Columbia, where I live now, decided tomorrow that it is going to build a nuclear plant. Or New Zealand, for that matter. How long would it take for the first watt of nuclear electricity to flow into that province's or country's grid?

Clearly not before a nuclear plant is completed. As the annual World Nuclear Industry Status Reports testify, the average nuclear plant takes around a decade to go from start of construction to producing electricity. (For the technically inclined, the weighted mean of the construction time for all reactors that became operational between 2011 and 2020 is 9.9 years, with construction time being defined as how long it took to go from when concrete was first poured at the base of the reactor and the reactor starts feeding electricity to the grid.)

To be sure, there is the occasional reactor that does get built in half that time, but these are compensated by those that take twice that time, which is why the average works out to be 9.9 years. And this time scale is despite favorable regulations and strong government support in most countries. For comparison, a utility-scale solar plant can feed electricity to the grid within about two years after building commences.

But that's only the physical construction time. One can't start construction of a nuclear reactor immediately. The requisite planning and fundraising—remember it costs billions to construct a plant—might take another decade.[5] Again, there are variations between countries in these time periods, but the total time period of around twenty years is typical.

Take the case of Finland. In May 2002, the Finnish parliament voted to build the country's fifth nuclear reactor; the following December, the power company Teollisuuden Voima decided to invest in an EPR unit. Construction of the Olkiluoto-3 reactor started on August 12, 2005, and it is only in March 2022 that the reactor was first connected to the grid, almost exactly twenty years after the parliamentary vote. Finland, however, is no novice in nuclear technology, having operated nuclear power plants since the 1970s. If it takes so long in Finland, then New Zealand or British Columbia are unlikely to do it any faster.

In the case of Hinkley Point C, the UK government's 2008 white paper that endorsed the idea of building nuclear plants to reduce emissions envisioned new reactors producing power by 2018. The white paper recommended choosing Hinkley Point as the location for the first nuclear plant because it already had the requisite environmental clearances. Despite that advantage, Hinkley Point C was nowhere near ready to produce power by 2018. Currently, EDF anticipates starting to generate power at this reactor sometime between 2029 and 2031.

The other advantage that should have helped Hinkley Point in being built on schedule and within budget was the United Kingdom's extensive experience with nuclear power. The country was an early leader in building nuclear power plants and over the decades has built forty-five of these—including one at Sizewell that is of a somewhat similar design to the EPR being constructed at Hinkley Point. Thus, one would expect that if nuclear plants could be built quickly, then the United Kingdom would be able to do so.

Experience with nuclear energy is not an advantage that many other countries have. The 2023 Status Report records that only thirty-six countries have ever operated nuclear power plants. Developing countries, in particular, have largely not built and operated them. But these countries are critical to the future course of carbon emissions, because they have mounting electricity requirements. How quickly these countries can build nuclear power plants will shape the technology's potential to contribute significantly to the mitigation of climate change. Therefore, a nuclear solution to climate change would involve many such countries acquiring their first nuclear power reactors and other facilities.

One of the few attempts at hypothetically laying out the geographical distribution of nuclear power in a scenario where it contributes significantly to climate mitigation was in an influential study published by the Massachusetts Institute of Technology in 2003. Its scenario envisioned the world having 1,500 gigawatts of nuclear energy capacity by 2050—over four times the 2023 nuclear generating capacity of 365 gigawatts—and foresaw a number of countries like Algeria, Indonesia, Malaysia, North Korea, the Philippines, Venezuela, and Vietnam all acquiring nuclear plants by 2050.

In this scenario, Indonesia, for example, would build up thirty-nine gigawatts of nuclear capacity by 2050. That means Indonesia should build around twenty-five large nuclear reactors like Hinkley Point C. Or if it were to think of lower power reactors, say small modular reactors, then it would need at least 130 reactors. Right now, Indonesia has none.

This absence is not due to a lack of interest. Successive Indonesian governments have been interested in nuclear power from the first decade of the republic's existence. In 1958, its first leader, President Sukarno, established the organization that was to eventually become the National Nuclear Power Agency (Badan Tenaga Nuklir Nasional, or BATAN). BATAN has been a consistent and powerful institutional driver for nuclear energy in Indonesia.[6]

BATAN's influence is most visible in the planning process.

Starting in the 1970s, Indonesian authorities have drawn up many plans. In addition to BATAN, the International Atomic Energy Agency has been another player. One national level seminar in 1974, for example, projected an installed nuclear capacity of 15 to 25 gigawatts by 2000; a 1976 IAEA report projected 6.4 to 20 gigawatts by 1992. None of that happened. Indonesian officials set some targets during this century too. In January 2006, for example, Indonesia's energy and mineral resources minister announced that the government would be calling for tenders for 4,000 megawatts of nuclear capacity, aiming for a completion date of 2016. That didn't happen either.

Same story in the case of Ghana. Ghana's interest in nuclear energy also dates back to its formation as a modern state.[7] Construction of its first experimental reactor started in 1964, with Kwame Nkrumah, the country's first president, inaugurating the project and singing praises of atomic energy. In the 1970s, the IAEA projected that Ghana would have 600 megawatts of installed nuclear power capacity by the end of the 1980s. That did not happen, but in 2007 another nuclear capacity target, this time of 400 megawatts by 2018, was approved by Ghana's government. As of 2024, there is not a single megawatt of nuclear capacity in Ghana.

Similar stories can be told of many other developing countries. But even in those that have built up nuclear plants, nuclear power remains an insignificant source of energy. India is a great example. The country is arguably far "ahead of the curve" when compared with most developing countries, when it comes to nuclear technology. Thanks to decades of sustained government support for the nuclear program, India's Department of Atomic Energy (DAE) has developed expertise and facilities that cover the entire nuclear fuel chain. Yet nuclear power contributes a mere 3 percent of the country's electricity, and it has remained at that level, or less, for decades.

Despite much hype about the potential role of nuclear power in delivering energy for development, the reality is that nuclear energy has not done anything of that sort. And this is likely to be the case going forward too.

There is a good reason why developing countries, despite a desire to build nuclear capacity, have not built nuclear power plants in large numbers. Financial resources for capital-intensive projects are scarce in cash-strapped nations, and nuclear plants are prohibitively expensive. And because of their emphasis on economic growth for alleviating poverty, these countries should prefer cheap sources of electricity generation. Nuclear power is an expensive and inefficient way to deliver energy to the developing world's unserved populations.

The potential time scale for expansion of nuclear power is important in determining whether nuclear power can contribute to climate change mitigation. As mentioned in the introduction, the Intergovernmental Panel on Climate Change and other international bodies have warned that to stop irreversible damage from climate change, net emissions have to be reduced drastically by 2030, and completely by 2050. Emissions from the electricity sector should be reduced even faster, simply because it is harder to decarbonize sectors like transportation, agriculture, and industry. Ideally, the electric grid will have to be nearly completely decarbonized well before 2050, especially in the Global North.

That time frame is a good yardstick to evaluate technologies positioning themselves as solutions. The potential role of an energy technology to mitigate climate change should be assessed on the basis of how fast the technology can be scaled up and how much emissions its deployment can save during this time frame. A decade is insufficient to even build one new nuclear plant from scratch.

Compare that with solar photovoltaics. Global capacity has gone from 72 gigawatts in 2011 to 707 gigawatts in 2020, a roughly tenfold increase, while the cost of generating solar energy in the United States has come down approximately by a factor of ten. China alone built solar power plants that can generate over 250 gigawatts (wind power plants, around 236 gigawatts) between 2011 and 2020.

The question that those concerned about the role of nuclear power in climate change should be asking themselves is not whether it is theoretically possible to build nuclear plants quickly. The important question is whether this can happen in the real world, in multiple countries as they go through being ruled by different political parties. Unless one simply wishes away the many problems that have historically afflicted nuclear projects, the answer is no.

This conclusion about nuclear energy being an infeasible option to tackle climate change typically attracts a host of counterarguments. In the following sections, I try to address some of the questions that I see come up often.

But nuclear plants can be made cheaper and built up faster, right?

The simple answer is no. There are two classes of reasons: historical experience and technical characteristics. The basic nature of nuclear power makes it inherently expensive and slow to build. This nature has been demonstrated repeatedly over the history of nuclear power around the world. This history is important. As Howard Zinn explained during a 2005 interview with *Democracy Now*: "If you don't know history it is as if you were born yesterday. And if you were born yesterday, anybody up there in a position of power can tell you anything, and you have no way of checking up on it."

History

Nuclear power plants have always been enormously expensive, even during the era of Atoms for Peace, with its "too cheap to meter" rhetoric. Those in the business of constructing power plants knew even then that nuclear power could be quite expensive.[8] As Lee Clarke documented in a 1985 paper in *Social Problems*, most US utilities were resistant to the idea of building atomic power plants and the federal government had to put a lot of pressure on them to persuade them to invest in these.

During the following decades, many utility companies were to find just how expensive nuclear power was as reactor project after reactor project ended up costing much more than estimated. In the United States, for the seventy-five nuclear plants whose construction started between 1966 and 1977, final costs and construction times exceeded initial projections by 207 percent and 94 percent respectively.

By the late 1980s, it was apparent that these reactors would be uneconomical.[9] Many of the projects were abandoned. In 2007, the US Congressional Research Service reported that "more than 120 reactor orders were ultimately canceled" within the United States. At a global level, France's Commissariat à l'énergie atomique et aux énergies alternatives (Alternative Energies and Atomic Energy Commission, or CEA) reported in 2002 that there were 253 "canceled orders" in thirty-one countries.

Some of these cancellations had larger repercussions. In August 1983, the Washington Public Power Supply System was involved in the largest municipal bond default in US history when it could not repay what it had borrowed to construct two nuclear power plants. The following year, the US Office of Technology Assessment opined that nuclear power "is an option that no electric utility would seriously consider." That prognosis held true for two decades, despite much effort by the Department of Energy, which tried to come up with various supposedly cheaper ways of building reactors.

It was only when the George W. Bush administration came to power that the nuclear industry's fortunes began to shift. The strongly pro-nuclear administration found many ways of siphoning public money into nuclear energy.

In 2001, a team led by Vice President Dick Cheney released a report titled, in truly Orwellian fashion, "Reliable, Affordable, and Environmentally Sound Energy for America's Future" that recommended, among other things, supporting "the expansion of nuclear energy"; just in case one might be wondering whether Cheney had a secret desire to address climate change, the report also recommended the promotion of "enhanced oil and gas recovery from existing wells through new technology." The Department of Energy, for its part, announced "A Roadmap to Deploy New Nuclear Power Plants in the United States by 2010."

All these reports were part of the drumroll about nuclear power that was all the rage during the first decade of this millennium, the so-called nuclear renaissance.[10] The promise there was that many reactors would be built.

To sweeten the deal for private companies, the US government introduced the Energy Policy Act of 2005. Among the provisions of the act that specifically applied to newly built nuclear reactors were funding for research and development, loan guarantees and insurance against regulatory delays, and a production tax credit. All were ways of making citizens cover financial risks and insure against any losses for utility companies. As a 2008 Congressional Budget Office report explained, "Loan guarantees and insurance against delays reduce the financial risk of investing in advanced nuclear power plants by transferring risk to the public" and even went on to add a cautionary note: "Economic theory suggests that such incentives cause recipients to invest in excessively risky projects because they do not bear all the cost of a project's failure."

Utility companies *did* invest in excessively risky projects. Altogether, they proposed building more than thirty reactors. Only four nuclear reactors proceeded to the construction stage, and two of these reactors were abandoned mid-project (see chapter 3). The remaining two are the ones at the Vogtle power plant in Georgia. In all these cases, the public bore the financial burdens resulting from the failure of these projects.

As the government rolled out these favorable policies, there was a burgeoning literature, extolling the much better economics of new nuclear reactors. Using a series of jargon-laden descriptions of new manufacturing techniques and seemingly sophisticated methodologies, they argued that these new reactors would be cheaper, perhaps much cheaper, than earlier reactors, and more generally, that nuclear power would be economically competitive.

These overly optimistic estimates projected construction costs of between $1 billion to $2 billion dollars for a 1,000-megawatt reactor. Such estimates came from nuclear reactor vendors, the US Department of Energy, and prestigious universities like the University of Chicago and the Massachusetts Institute of Technology. The low projections for the cost of construction were matched by their assumption that these reactors could be constructed in three to four years.

One reactor design prominent in this hype was Westinghouse's AP1000 design. When Westinghouse started promoting this design, it claimed that it would be built for about $4 billion (in 2020 dollars) within thirty-six months. This was the reactor design chosen for the Vogtle nuclear project in Georgia, which ended up at around $35 billion for two such reactors and with a construction time of over ten years.

Technical attributes

At the most fundamental level, the reason for this high cost is that a nuclear power plant is just a very elaborate way of converting water to steam. But the underlying process used to generate the heat that is used to boil water—namely, nuclear fission—is inherently hazardous. Thus, controlling this hazardous process ends up requiring a very contrived technological artifact, somewhat reminiscent of a Rube Goldberg machine.

Cultural theorist and literary critic Sianne Ngai offers this apt description:

> Starting in the late 1920s, Goldberg drew hundreds of devices he called "inventions," machines ostensibly supposed to help people save time but that were in fact elaborate contraptions involving lots of parts to accomplish a simple task ... What's put on display is an excessive amount of effort to achieve a minimal result.[11]

In a similar vein, a nuclear power plant represents an excessive amount of temporal and financial investment to produce electricity, which could be produced using much less investment. Because of the sheer complexity of the process, it will always be expensive. Just as no amount of innovation can lower the cost of a battleship to that of a simple canoe, technologies like photovoltaics will always be cheaper than nuclear plants. Provided, of course, one is interested only in electricity and not in making nuclear weapons (see chapter 5).

Focusing on building nuclear plants or fossil fuel plants is inefficient in a different sense too. This focus loses sight of the reason to build power plants, which is to provide services that people demand—for example, illumination or heating. More efficient means of providing these services could be far better economically and environmentally. Even better would be to do away altogether with demands that are unnecessary for well-being—energy used to produce cryptocurrency, for example. More efficient energy planning to reduce demand was one of the central ideas motivating the Seoul Metropolitan Government's "One Less Nuclear Power Plant" program that Mycle and nine other energy analysts advised.

But let us return to nuclear power plants and why they are expensive. Not only are nuclear plants complicated like Rube Goldberg devices, they can also experience catastrophic accidents. This unique characteristic necessitates the multiple safety features used in a nuclear plant, which in turn require vast quantities of materials. Also required are workers with high levels of training, necessitating higher salaries, and a regulatory infrastructure to ensure that plants operate safely. All of these compel expending lots of money.

Nuclear power promoters don't like to admit to these underlying characteristics. Instead, they blame excessive regulation for the high costs. This argument has become especially resonant in recent decades, thanks to neoliberal antipathy towards regulation.

The free-market-supporting think tank Institute for Energy Research puts it bluntly, if unsurprisingly: "Regulatory burdens on nuclear plants are making them expensive." Steve Kidd, who worked for seventeen years with the lobbying organization World Nuclear Association, argued that nuclear advocates should concentrate on undoing what he termed the tangle of regulations.

One more illustration to prove the point. In a 2019 episode of the podcast *Titans of Nuclear* (in which proponents of nuclear technology are invited to sing its praises), the host Bret Kugelmass asked an official from GE Hitachi: "And why not just build [a nuclear reactor] in Uganda, where they don't even have a regulator? ... Why not just skip the whole regulatory hassle [and] build the same thing you would have built here but just ... in Uganda [and] not have the billion-dollar price tag?" In other words, Kugelmass's strategy to build cheap is to skip any regulatory oversight whatsoever.[12]

Skipping regulatory oversight with the technology that brought us Chernobyl is a fantastic idea. Provided one could follow the author Terry Pratchett's advice: "Eight years involved with the nuclear industry have taught me that when nothing can possibly go wrong and every avenue has been covered, then is the time to buy a house on the next continent." Just kidding, of course.

Even without thinking about countries without any nuclear power plants, such as Uganda, regulation is too lax as it is, with a variety of structural and political factors inhibiting effective oversight of the industry.[13] In turn, this is the source of various small accidents as well as potentially large ones—the near miss at the Davis-Besse plant is a good example of what could happen when regulators allow the industry to self-regulate.

In 2002, leaking boric acid almost ate through the steel in a key part of a nuclear reactor—the pressure vessel head—at the Davis-Besse nuclear plant in the state of Ohio. The result was a gaping hole the size of a football. There remained only a thin stainless-steel lining that protected the nuclear reactor from "a meltdown with a large release of radiation to the atmosphere."

There had been indications earlier, such as dried boric acid deposits on the outside of the vessel. But oversight was lax and routine inspections had failed to detect the ongoing process of corrosion. Just prior to the discovery of the hole, the Davis-Besse plant received the highest ratings possible in the US Nuclear Regulatory Commission's Reactor Oversight Process.

The plant was to be shut down by December 2001 for a full inspection, but the operating organization got NRC's approval to postpone full inspection by some months. The NRC's inspector general attributed this approval "in large part" to wanting to lessen the financial impact on the utility. But for sheer luck, the financial impact on the utility and the taxpayers of the country could have been far worse.

While Davis-Besse was a near miss, the accident that did happen was Fukushima. The roots of that disaster have been traced to weaknesses in the Japanese regulatory system. The Rebuild Japan Initiative Foundation's Independent Investigation Commission, for example, identified "the sweetheart relationships and revolving door that connected the regulatory bodies and electric companies, academics, and other stakeholders in the nuclear community" as one cause of the Fukushima accident.

The underlying problem goes back to the US regulatory system. Even in the 1970s, General Electric's BWR Mark I design was known to be incapable of withstanding a severe accident. Behind the scenes at the US Atomic Energy Commission, analyst Stephen Hanauer had recommended a stop to licensing that reactor design and other designs that had the same weakness. His idea was dismissed; the commission wanted nothing to come in the way of continued nuclear reactor construction. A senior official argued that stopping the licensing of GE plants "would generally create more turmoil than I can stand thinking about" and thus might spell "the end of nuclear power."[14] The price for such decisions was paid by the people of the Fukushima region.

As long as nuclear power plants operate, they need to be strictly regulated. The alternative is a higher risk of accidents that could result in widespread and long-lived radioactive contamination, not to mention very high financial costs for cleaning up.

Can't new manufacturing methods lower costs?

The nuclear industry claims to have strategies to reduce construction costs and times. Most prominent among these is the use of what is termed modular construction. The idea is to manufacture many components of the reactor in factories and put them together on the site. But modular construction is not new. It has become a standard practice in modern manufacture of many goods. Today's apartment complexes, for example, are not built brick by brick but by assembling different factory-made components: walls, windows, and so on. Nuclear reactors, too, have been built using modular construction methods.

The most recent example is the leading US nuclear vendor Westinghouse. Its AP1000 reactor, the one being constructed in Georgia, features modular construction. Westinghouse officials regularly hailed this approach as a "significant innovation." Belief in modular construction was the basis of Westinghouse's projection that the AP1000 reactor would take three years to build. And that it would cost the $2 billion figure mentioned earlier. These were then recycled by lobbyists —for example, by the Nuclear Energy Institute—in congressional testimony.

All those prognoses proved completely wrong. Every one of the AP1000 reactors built so far, whether in the United States or in China, were all significantly delayed and cost more, as detailed earlier. A former member of the Georgia Public Service Commission, the state utility authority overseeing the Vogtle nuclear power plant, summed it up aptly to the *Wall Street Journal* in July 2015: "Modular construction has not worked out to be the solution that the utilities promised." People in Georgia should know: the AP1000s at Vogtle are proving to be many times more expensive than what was first promised by Westinghouse.

Rather than learning from this debacle, the nuclear industry has doubled down on the claim that modular construction will lower costs. Hence the "modular" in small modular reactors, the latest offering from the industry. This does not bode well for the nuclear industry's attempts at lowering costs through innovation. Those interested in evaluating the economics of nuclear power should be deeply skeptical of such claims.

But the nuclear industry can learn from its experiences, right?

Another argument advanced by nuclear advocates is that they can lower costs by "learning." Again, the evidence so far suggests that this is unlikely. A recent examination by a group from the Massachusetts Institute of Technology showed that in fact costs had increased with time. Their analysis shows that as total nuclear capacity doubled, say, from 50 gigawatts to 100 gigawatts, the cost of a new reactor would be 2.15 times the cost of a reactor built at the earlier stage.

From a nuclear advocate's point of view, the United States might be seen as an example best not to emulate. Even though it has the largest fleet of nuclear power plants, it has not standardized its reactors, in turn because it has too many different actors to be able to build on earlier experiences. Adding to this is the charge of mismanagement, with the conservative *Forbes* magazine declaring in 1985 that the US nuclear power program "ranks as the largest managerial disaster in business history."

The French, on the other hand, are the darlings of the pro-nuclear crowd. It has one of the highest nuclear shares of any country. It adopted relatively standardized designs. Policymaking has been tightly controlled by the government and a couple of large national institutions. So if learning were to work anywhere, it should work in France.

And that is what the energy analysis community assumed for decades—till Arnulf Grubler, an engineer at the International Institute for Applied Systems Analysis, actually examined the cost figures. Using official French government sources, he showed that in comparison with the set of six reactors built between 1971 and 1979, the four reactors built between 1984 and 1999 were over double the cost and took twice as long to construct. Per unit costs rose by a factor of 2.5 over the period he studied, despite "a most favorable setting"; more generally, he concluded "nuclear reactors across all countries with significant [programs] invariably exhibit negative learning, that is, cost increase rather than decline."

Globally, too, there is a great deal of evidence for increasing costs. Some point to specialized conditions—specific countries, time periods, and so on—when costs have gone down. But these are not generalizable, and unlikely to be reproduced everywhere. That makes their applicability dubious at best when discussing a global expansion of nuclear power. In other words, any serious discussion of nuclear power as a solution to climate change should start with the clear understanding that reactors will take a long time to commission and be very expensive, and the costs will likely rise with time.

The nuclear industry has had decades to try and establish itself as an economical source of energy, and it has failed. The continued claims that the next round of reactors will be cheap and built on time and within budget are akin to a repeated theme in the *Peanuts* comic strip. The theme involves Lucy inviting Charlie Brown to kick a football that she is holding, only to remove the ball at the last second, causing Charlie to fall. As readers of *Peanuts*, we learn not to trust Lucy and her assertions that this time it will be different. But for a variety of reasons that we explore in the rest of the book, policymakers and some investors continue to fall victim to the nuclear industry's Lucy-like claims. The urgency of the climate crisis means that we have neither the financial resources nor the time to be engaging in this game.

The infeasibility of using nuclear power to mitigate climate change

Investing in expensive nuclear power is inefficient and ineffective. With his characteristic emphasis on efficiency, Mycle Schneider told *Deutsche Welle* that "every euro invested in new nuclear power plants makes the climate crisis worse because now this money cannot be used to invest in efficient climate protection options."

Efficient climate protection options—like renewables, energy efficiency, and demand response, to name just a few—cost far less than nuclear reactors and will reduce the carbon dioxide emitted far more quickly. Expanding nuclear energy can play only a negative role in mitigating climate change.

3

Private Profits, Social Costs: Industry Strategies

The object of Fisker, Montague and Montague was not to make a railway to Veracruz, but to float a company. Paul thought that Mr. Fisker seemed to be indifferent whether the railway should ever be constructed or not. It was clearly his idea that fortunes were to be made out of the concern before a spadeful of earth had been moved.

Anthony Trollope, *The Way We Live Now*

Finding new ways to privatize the commons and profit from disaster is what our current system is built to do; left to its own devices, it is capable of nothing else.

Naomi Klein, *This Changes Everything: Capitalism vs. the Climate*

Kevin Marsh's passion was cars, especially fast ones. He inherited the fondness for cars from his grandfather, an auto mechanic. Once Marsh had become wealthy by rising through the ranks of the electric utility industry, he splurged on these. In 2017, he owned ten cars worth over $800,000, reported the *Post and Courier*. The fleet included a nearly $300,000 Aston Martin Vanquish and a $117,000 BMW M6.

Marsh's other interest was helping the less fortunate. He liked serving on the boards of organizations that don't show up on the front page of the newspaper. Because "they're the ones that really need the help," he once explained. Marsh acquired this desire to help society from his mother, who worked with various community organizations in Athens, Georgia. For eight years, Marsh served on the board of the Epworth Children's Home, a residential facility for children from broken homes. Speaking to the *Columbia Metropolitan*, the organization's president described Marsh as "a dedicated board member" who "brings a big heart" and genuine care "for children and for people who maybe don't have it as good as a lot of other folks do."

All that charity work helped Marsh, especially after he pleaded guilty to conspiracy to commit mail and wire fraud. AP News reported that his lawyers submitted ten letters detailing Marsh's record as a do-gooder "from helping the family of an employee killed on the job get financial and legal help to securing an air conditioner for a women's home and taking a week out of his busy executive schedule to volunteer for vacation Bible school." One doesn't know whether these letters influenced the judges, but Marsh received just a two-year sentence.

Kevin Marsh's fraudulent activities were committed while trying to build a nuclear power plant in South Carolina. As is the case with most nuclear projects, Marsh and his coconspirators made customers in the state pay a large fraction of the enormous costs involved. They were also counting on a couple of billions from US taxpayers. But in a familiar twist, the project became delayed and the cost estimates rose. At some point, Marsh and his buddies realized that even these twin cash cows, ratepayers and taxpayers, couldn't have forked over enough money to keep this charade going, and the project was abandoned. That's when Marsh's troubles began.

Digging into the history of a failed project reveals much about nuclear power. In particular, it offers a partial answer to a question often posed to me when I discuss the economics of nuclear power: If nuclear power is as expensive as I described in chapter 2 and it takes so long to build a reactor, why do corporations get involved in this enterprise at all? The answer is complicated, but the simplified version is that they do so only when the public can be made to bear a large fraction of the high costs of building nuclear plants and operating them, either in the form of higher power bills or in the form of taxes. Then many companies find nuclear power attractive.

Nuclear infrastructure is expensive. Reactors, geological repositories, uranium-enrichment plants, you name it—everything costs a lot to build. But high cost is far from being a negative trait, for companies benefit from large contracts and huge profits.

Indeed, much of the money is made well before any of these facilities are completed. As of 2008, *World Nuclear News* reported that the Yucca Mountain project had cost about $13.5 billion, split eighty-twenty between electricity consumers ("ratepayers") and citizens ("taxpayers"). That project is essentially dead, despite the effort spent by the nuclear industry and some politicians on keeping it alive. Not a single ounce of waste will likely be buried there, ever. But for over four decades, the proposed repository has been the gift that kept giving, to a large number of consultants, corporations, and national laboratories.

The prerequisite for all those profits is to find a means to make the public pay. For state-owned companies—Chinese or Indian nuclear plant operators, for example—this poses no major problems. But companies in the United States or the United Kingdom have to work harder. In this chapter, I detail how a US company made money off a nuclear plant that never produced anything useful to consumers, and how another company is planning to profit from buying up nuclear plants that have stopped operating. I will also briefly discuss how many of these organizations ally with others to build up political clout. These mechanisms explain why the nuclear industry continues to be interested in investing in a source of energy that is evidently uneconomical from the viewpoint of consumers. These methods also play a big role in convincing governments to allow them to socialize their costs and privatize their profits.

Such methods are not unique to the nuclear industry. These have become common practices under capitalism today, which involves a small number of organizations wielding inordinate amounts of economic and political power. Understanding how the system works to benefit these powerful players will help us also figure out how to transition our society into one that is compatible with ecological constraints. Conversely, a system that allows such players as the nuclear industry to continue to profit will not permit rapid and wide-ranging action on the climate.

A nuclear cluster in South Carolina?

Kevin Marsh's professional career started with a stint at the accounting firm Deloitte & Touche. He then joined South Carolina Electric & Gas Company (SCE&G) in 1984. SCE&G is an investor-owned utility company that dates back to 1924, when it was called Broad River Power Company. Over the decades, it had expanded by acquiring multiple companies. The same year that Marsh joined, the company merged with Carolina Energies to form SCANA, a holding corporation.[1] SCANA itself was acquired in 2019 by Dominion Energy, an even larger power company with a market capitalization of over $60 billion at that time. Such growth through mergers and acquisitions has, over the decades, led to the electricity industry becoming an oligopoly with a few companies like Dominion dominating US energy supply.

Marsh rose through the ranks, and in 1996 he became the senior vice president and chief financial officer of SCANA, then in 2006, the president and chief operating officer of SCE&G. The CEO of SCANA at that time, Bill Timmerman, was a cheerleader for nuclear power. He drew on the ideas of Michael Porter, a professor at the Harvard Business School, to promote the concept of a "nuclear cluster" in South Carolina. And as

Timmerman groomed Marsh to take over after him as CEO of SCANA, Marsh, too, inherited his mentor's nuclear dream.

South Carolina was among the first states in the country to host a nuclear reactor. Nuclear fission took root in 1953, when the R Reactor, used to produce plutonium for the US nuclear weapons arsenal, went critical at the Savannah River site. As of January 2022, the state hosted seven nuclear reactors, the highest on a per capita basis. Only Illinois and Pennsylvania hosted more reactors, but they are far more populous. Six of South Carolina's reactors were operated by Duke Energy, an $80 billion company and a political heavyweight in the state. This lengthy history and concentration of nuclear reactors meant that the state's economic and political elite would be—or could become—interested in building a "nuclear cluster."

The first step in establishing a cluster, according to SCANA, was to build two AP1000 nuclear reactors. (Duke, for its part, planned to build two AP1000 reactors at the Lee nuclear station, but eventually scrapped the idea after spending over half a billion dollars.) SCANA's plant was to come up in Jenkinsville, about twenty-five miles from the capital, Columbia. The site already housed the V.C. Summer nuclear reactor, which had been operating since 1984. That reactor was owned jointly by SCE&G and Santee Cooper, South Carolina's state-owned company set up during the New Deal in the 1930s. The two firms planned to co-own the proposed two reactor unit, too, with SCE&G taking the majority stake.

In 2007, Santee Cooper and SCE&G joined hands with Duke Energy and another large electricity company, Progress Energy, to get South Carolina's legislature to pass the key catalyst for the proposed nuclear plant: the Base Load Review Act. The legislation described itself as an "act to protect South Carolina ratepayers," but could be described better as an act to protect corporate and investor interests. It was so egregious that even the governor, Mark Sanford, a Republican, was opposed to the act, because it violated his free market ideology—he "wanted nuclear power to rise or fall without government help"—but he could not veto the bill, since the legislature would override it. Drafted with input from an attorney who worked with SCE&G, the Base Load Review Act read like a wish list that corporate executives would have dreamed up. Most importantly, it enabled the South Carolina Public Service Commission to authorize the utility to charge customers for costs involved in the development of the proposed nuclear reactors. What could be charged included "evaluation, design, engineering, environmental and geotechnical analysis and permitting, contracting, other required permitting including early site permitting and combined operating license permitting, and initial site preparation costs and related consulting and professional costs"—in other words, just about everything. Crucially, it allowed SCE&G to pass on the interest and other financing costs for the multibillion-dollar loans it needed to proceed with the project.

Transmitting costs to customers is not new for electricity firms. Customers have historically paid the costs associated with power plants, plus a profit margin for the company. What the Base Load Review Act enabled was having customers pay these well before the nuclear plant started generating power. As a result, even though they were never to get any electricity from this facility, South Carolina customers found their monthly bills go up by about twenty-seven dollars, according to an in-depth article in *The State*, though that amount is being fought out in the legislature and the courts.

The pretense on the part of SCE&G officials was that these higher bills would help consumers. In a YouTube video posted by the Nuclear Energy Institute, Steve Byrne, Marsh's deputy, argued,

> Recovering it as you go as opposed to accumulating allowance for funds used during construction means we can actually lower the interest charges to our customers ... We estimate that it will

save about a billion dollars over the construction phase and over the entire life of the project and even after it goes into service will save our customers about four billion dollars.[2]

One might be forgiven for thinking that people like Byrne and Marsh cared deeply about the best interests of their customers.

Making the argument for a nuclear plant

Prior to charging its consumers, SCE&G had to persuade the South Carolina Public Service Commission that building two large nuclear reactors was a prudent choice. In May 2008, Kevin Marsh wrote to the Public Service Commission and requested a "Certificate of Environmental Compatibility and Public Convenience and Necessity and Base Load Review Order" and to "authorize the Company to put into effect the rates" viewed by SCE&G as necessary for profitability.

SCE&G's argument for why the V.C. Summer project was "prudent" followed a boiler-plate pattern used by nuclear proponents around the world to justify reactors: unless they built these nuclear plants, SCE&G could not reliably provide its customers electricity in the future. To borrow from chess terminology, one might term this argument for nuclear power the "lights will go out" gambit.

Marsh's testimony to the commission in December 2008 elaborated the necessity argument. After describing various drivers of SCE&G's energy sales, Marsh detailed their energy-demand forecasts. Utility companies regularly produce these forecasts as part of something called the Integrated Resource Plan. In South Carolina, this plan is updated every year and presented to the Public Service Commission every three years.

These exercises are meant to induce utility companies into making the best choices for future energy supplies. Like many such good ideas, the exercise doesn't produce the desired outcome, because corporations have figured out how to manipulate this process. The typical result is the triumph of private profits over the public interest.

One way corporations game this process is by exaggerating their need for new power plants. Overestimating demand and building more plants benefit companies like SCE&G, which operate within a market structure where the price of electricity for customers is set to compensate the plant owner for all expenditures plus a fixed percentage of those expenditures as profit.

A widely cited paper from 1962 by two economists, Harvey Averch and Leland Johnson, identified the problem with this arrangement: "The firm has an incentive to acquire additional capital if the allowable rate of return exceeds the cost of capital." Translation: if a utility company could borrow money and invest in a project and charge its consumers enough to pay off those interest charges, then it would be tempted to incur such expenditure, regardless of whether consumers need this project.

The only restriction for SCE&G and other such companies was that the project expense should be deemed allowable, which meant that the utility had to convince some government regulatory body that it was a necessary investment. This is where the exaggerated demand estimates and the "lights will go out" gambit comes in. Once the regulator has approved investment in a nuclear power plant, or any other generating facility for that matter, then electricity consumers end up having to pay whatever costs accrue from that project.

The high cost of constructing a nuclear plant, then, is not a problem for the project developers, because they get paid in any case. Indeed, the more the cost, the more the profit to the utility, albeit at the expense of consumers, who will pay more for their electricity. Utilities, therefore, regularly gold-plated capital additions to earn more profit, as a 2003 paper in the *International Journal of Management* explains.

SCE&G, then, had an economic incentive to posit the need

for new power plants. And it did precisely that in the 2006 Integrated Resource Plan that formed the basis of Marsh's 2008 testimony to the commission. In this plan, SCE&G claimed that its energy sales would increase by 22 percent between 2006 and 2016, and by nearly 30 percent by 2019.

In fact, SCE&G's energy sales *declined* by 3 percent by the time 2016 rolled around. But that was for the future. In 2008, this never-to-occur increase in energy demand was the primary rationale advanced by Marsh to the South Carolina Public Service Commission.

Having made this argument about a shortfall in supply, the next step was to make the case that the two AP1000 reactors represented the most appropriate way to meet this projected demand in 2016 and 2019. Marsh's testimony, therefore, went on to describe how diligently SCE&G had done its research on the costs of nuclear power plants and other options. The AP1000 reactors, he told the commission, were "clearly … best suited for SCE&G's needs." He particularly praised the two companies that were to supply the nuclear reactor design (Westinghouse) and construct the reactors (Stone & Webster) and described in detail how SCE&G had negotiated a great contract with these companies.

Based on the information provided by Westinghouse and Stone & Webster, SCE&G initially projected the construction cost of the two nuclear reactors at $4.94 billion (in 2006 dollars, which is roughly $6.23 billion in 2020 dollars). That cost quickly doubled to $9.83 billion even before the full licensing application was submitted. Of this, SCE&G was responsible for 55 percent of the costs; the remaining 45 percent would be covered by Santee Cooper, in exchange for which it would receive the same percentage of electricity output. SCE&G was to be the formal operator of the plant.

Missing representation for the public interest

These arguments from SCE&G for constructing two nuclear reactors were full of holes and a conscientious regulator looking out for the public's interest would have rejected the proposal. The commission did not even have to analyze the application in great detail; it could simply have drawn upon the counterarguments laid out by nongovernmental and civil society groups. Tom Clements from the environmental group Friends of the Earth pointed out in the *Bulletin of the Atomic Scientists* in 2021 that most of these groups proved "prescient in their early assessments of the project."

Clements is a local activist who has fought the nuclear craziness in South Carolina for decades. Known for often using eye-catching props—at one rally, he held aloft a giant check for "limitless billions" made out to the SCE&G—he is one of the few citizens that consistently questioned the commission's greenlighting the V.C. Summer project and passing on costs to consumers. This consistent opposition did not endear him to the commission or SCE&G or Westinghouse. As he emailed me in 2012, "The nuclear industry wants total control and no discussion except on their terms. All the more reason for me to keep up asking questions and [not] accepting the status quo."

When Friends of the Earth intervened before the Public Service Commission in August 2008, they argued that SCE&G's application should be denied. The group emphasized that the "cost of the reactor project will be astronomical and ... likely to spiral out of control"; they also pointed out that SCE&G had provided "almost no analysis of the use of alternative sources of power"—in particular, wind and solar energy—as well as energy conservation and efficiency.

Both points were painfully obvious, but not to the Public Service Commission, which approved the company's request. The decision was, as Clements characterized it to the *Associated Press*, "a clear sell-out of the public interest over the interests of SCE&G." It was also not the last time that the commission chose the corporate over the public interest at V.C. Summer. Years later, Clements would exhort the commission to side "for once" with customers and require that SCE&G and its shareholders be forced to "bear a major portion of the cost increase." To no avail.

Shortly after the 2009 decision, Friends of the Earth legally challenged the Public Service Commission's approval, but the state's Supreme Court ruled for the commission. Clements also joined activists around the country to try and stop the federal nuclear subsidy programs, but that failed as well. The organs of political power unambiguously favored the corporations.

Dipping into the federal trough

In addition to South Carolina customers, SCE&G also relied on US taxpayers. As part of President Bush's nuclear power advocacy, his administration had enacted the 2005 Energy Policy Act (see chapter 2). That offered several incentives, including tax credits that companies can receive in exchange for each unit of electricity that is produced in their plants. These credits were valued at approximately $2.2 billion for V.C. Summer. The catch: the tax credits were available only if reactors started generating power by January 1, 2021.

When they were announced, the nuclear industry and its proponents were happy with the tax incentives. But like Oliver Twist, they wanted more. The Nuclear Energy Institute approached Congress and lobbied for a 30 percent tax credit for just investing in building a new nuclear reactor instead of waiting until it produced electricity; as a backup, they suggested the deadline be pushed back to the start of 2025, according a 2014 Congressional Research Service report. In other words, even as nuclear advocates were advertising how quickly a new generation of reactors would be built, they were unhappy about limiting possible tax credits only to plants that actually generate power and that do so within fifteen years. They preferred getting their rewards for just investing capital, even if that money produced nothing of use—like electricity— to the public. Unfortunately for SCE&G's executives, that attempt failed, and those restrictions remained in place. Publicly, SCE&G appeared confident about meeting the deadline; indeed, the financial projection submitted to the commission in 2008 assumed tax credits starting from 2016.

The project splutters

Construction of the reactors started in March 2013. After the concrete was poured to make the base for the reactor—traditionally, the marker for the official start of construction—Kevin Marsh called it an "exciting achievement." Later that year, he told *Columbia Metropolitan* about his belief that "customers will be served with that energy source 60 years down the road" and that "it will be a huge benefit to them to have clean, reliable, safe energy."

Highlighting his great concern for customers, he told the interviewer how he sometimes puts "on headphones" and listens to customer complaints. "It was heartbreaking to hear the stories that people have and some of the challenges they've run into that impact their ability to pay the bill ... It's cold outside, you don't have heat and don't have electricity; it can be scary. They don't want to complain because it's high or they think they've been overcharged. They are asking, 'Help me figure out how to pay my bill.'" The logic of this statement is perplexing: increasing the customers' bills cannot make it easier for them to pay the bill.

What Marsh wasn't telling his interviewers was that the project was delayed. There was a pattern to these delays. During most years after 2009, SCE&G would go back to the Public Services Commission and petition for revisions to the schedules and costs, and the commission would approve. This happened in 2010, 2011, 2012, and 2015. By 2016, SC&G was projecting a cost increase of 51 percent, from $4.5 billion to $6.8 billion (in 2007 dollars) for *its share* of the project. Overall, the project cost was being reported as $16 billion by 2017.

The problem was at the other end, and a familiar one at that. Westinghouse and Stone & Webster could simply not live up to their promise of building reactors at the quoted cost or timeline. And there was plenty of blame to spread all around.

Westinghouse's design had a number of shortcomings, but these became apparent only after construction started. Most of these problems can be traced to the reactor design—which nuclear advocates described using glowing adjectives like "innovative" and "novel"—and its supposed virtue: modular construction. Building a reactor in a modular fashion involves two sets of activities. First, a factory would fabricate parts (or modules) of the reactor and ship them to the construction site. Second, these prefabricated parts would be assembled together at the site to make the nuclear power plant.

One can think of these modules as being somewhat like Lego blocks. For some years, Westinghouse's website even featured a three-minute animated video that showed week-by-week progress; each week, one or more of these Lego-like modules would be tacked on and the AP1000 reactor would be completed in 116 weeks. It looked very smooth and exciting.

Except it didn't work that way in South Carolina, or at any of the other sites for AP1000 reactor projects. To start with, the AP1000 design was far from complete. Over the course of constructing the V.C. Summer and Vogtle projects, Reuters reported in May 2017, Westinghouse made "several thousand" technical and design changes.

Once the project had been canceled, a SCE&G official acknowledged at a State Senate hearing that it never received "a fully integrated schedule" from Westinghouse. But this should have been expected. Even before the V.C. Summer project commenced, Westinghouse's AP1000 design had been selected by China's State Nuclear Power Technology Corporation. Anyone following the saga of AP1000 reactors in China—and one would expect people like Marsh to have been paying close attention—the problems with Westinghouse's product should have been embarrassingly obvious. At the Sanmen nuclear plant, a layer of shielding that surrounded the reactor pressure vessel expanded and seeped out. That created such a mess that it prompted a Westinghouse official to admit to the *Wall Street Journal* that it was "disappointing." The steel supports for the 115-ton pressurizer, which helps control pressure levels, were too weak. Such problems kept recurring, demonstrating the folly of assuming that pretty computer-generated graphic simulations would actually work in the real world.

The next round of problems occurred while converting the "final" design into actual equipment and a fully constructed reactor. Plans simply did not work the way they were supposed to. Workers, it turned out, could make mistakes. And managers could cut corners or be deceptive. Such problems are not unique to nuclear manufacture. But the consequences of manufacturing errors could be a devastating accident leading to large-scale radioactive contamination, and that is unique to nuclear power.

Let us start with an example from the Lake Charles facility in Louisiana, where some of the components of the V.C. Summer reactor were being manufactured. In 2013, workers dropped and damaged a prefabricated building section and then tried to cover up the accident. After a lengthy investigation, the Nuclear Regulatory Commission concluded that "a former company official deliberately instructed subordinates to initially provide false statements as to the cause of the drop" reported *Nuclear Intelligence Weekly*. The NRC also found that the manufacturers had improperly labeled components, or had produced parts with wrong dimensions, and neglected required tests.

All these failures during manufacture could be traced to deficient quality control. Companies involved seemed "clueless" about the complexities involved in the manufacturing process, such as welding for nuclear reactor components, according to *Engineering News-Record*. Worse, the managers in charge of the manufacture seemed hostile to the idea of maintaining high standards in production and complying with regulations. So much so that a senior manager threw a letter opener at a junior official expressing concern about these problems and pushing the company to pause work and fix them.

There were similar errors at the construction site in South Carolina. In 2015, workers were drilling into concrete when they went too far and damaged the containment vessel, a component critical to the safety of the reactor. In addition to individual workers making mistakes, there were problems with planning too. SCE&G's schedule had not carefully considered the details of putting together the different modules of the reactor. It turned out, for example, that something called the CA03 module could be installed only *after* the so-called CA01 module had been installed. This meant that the delays in one added to the other. In 2014, SCE&G concluded that the reactor would likely be ready two years later than originally envisioned—which would be a problem because of the production tax credit requirements.

While these problems and delays and consequent cost increases were mounting, the two major corporate entities involved at the back end became involved in a series of lawsuits and counter-lawsuits—a brawl, as the *Economist* bluntly characterized it. Ultimately, in 2016, Westinghouse purchased the nuclear construction unit Stone & Webster to avoid these lawsuits. The hope was that this would speed up construction. That ploy failed. But in the process, Westinghouse ended up becoming liable for the cost increases at the reactors being built.

The following year, on March 29, 2017, Westinghouse, the largest historic builder of nuclear power plants in the world, filed for chapter 11 bankruptcy protection. The *New York Times* called it a "blow to nuclear power" and pointed out that it was companies like SCE&G that would find it hard, because they had to absorb losses that Westinghouse could not cover.

Fallout in South Carolina

Back in South Carolina, Kevin Marsh was determined to stay the course. During a conference call with investors immediately after the Westinghouse announcement, he stated, "Our commitment is still to try to finish these plants. That would be my preferred option. The least preferred option, I think realistically, is abandonment." By then the state had a new governor, Henry McMaster, who was very supportive of the project. This governor put out a statement expressing the hope that the project would be completed and declaring confidently that SCE&G and Santee Cooper must have prepared "plans and contingencies" that would "result in the completion of the project." In fact, the main contingency plan for SCE&G, reported the *Post and Courier*, was to hire a bankruptcy attorney out of concern for Westinghouse's financial status.

Marsh's stated hope was that Westinghouse would cough up enough of a compensation for what it hadn't delivered to cover additional costs. That was patently unrealistic. After all, Westinghouse had sought bankruptcy protection precisely to avoid such payments. In any case, the additional costs had swollen so much that even a solvent Westinghouse could not have filled the gap sufficiently and South Carolina's electricity consumers would have to pick up even more of the tab than they already had.

Santee Cooper, on the other hand, evaluated how the project was going and concluded that if it went ahead with construction, it would have to spend 75 percent more than it had budgeted, and the two units would be over four years late. Four years late meant that the project would not qualify for the federal tax credits. In July 2017, Santee Cooper announced that its board had voted to suspend construction, and SCANA reluctantly followed suit. Ceasing work, Marsh told the *New York Times*, "was our least desired option, but this is the right thing to do at this time."

But there was money to be made here too. The Base Load Review Act had generously allowed SCE&G to recoup even the "abandonment costs estimated at $4.9 billion" at a "guaranteed rate of return" of 10.25 percent, and the company could be receiving money for "its costly mistake" according to *Nuclear Intelligence Weekly*. If that were spread out over the six decades that the BLRA allowed, then SCE&G's customers would be paying the company hundreds of millions of dollars for that entire period. At that time, 18 percent of their monthly electricity bills already went to the abandoned nuclear reactors and associated costs.

Further, SCE&G had issued around $3.5 billion in long-term corporate bonds to finance the project. These bond holders had to be paid their financing costs, roughly a figure of $2 billion, as of 2018. In other words, the expenditure on the project has continued to increase even after it was abandoned. Customers continue to be charged for this component of costs.

South Carolina customers got a small bit back when they launched a class action lawsuit against Dominion Energy, which had purchased SCE&G. In 2019, Dominion agreed to pay $60 million and another $61 million in 2022. In the first tranche of payments, the resultant checks that customers received in the mail were so small—some as little as 4 cents—that many did not even bother to cash those checks; more than 10 percent of the money went unclaimed.

In all, at least $9 billion had been spent on construction, roughly $5 billion by SCE&G and $4 billion by Santee Cooper.

All that was left to show for that expenditure was a big hole in the ground.

South Carolina is not the only instance where customers had to pay for projects that never materialized. In Florida, for example, customers paid nearly $900 million for the canceled Levy County units. To companies in the business, nuclear reactor construction is truly a gift that keeps on giving, while robbing society at large.

None of this bothered South Carolina's political elite—till the V.C. Summer project was abandoned. Then the state's House of Representatives and Senate set up committees to investigate what happened and conducted legislative hearings. Kevin Marsh, Steve Byrne, and other SCANA officials had to testify. Suddenly, even the Republicans found the actions of SCE&G problematic. Peter McCoy, who was then the chair of the House Utility Ratepayer Protection Committee and who was appointed by President Trump to the position of US attorney, seemed appalled that the ratepayers had to make up SCE&G's losses while investors continue to "make a 10.25% return."

That was only one element in SCANA's troubles. In parallel, the federal government and the US Securities and Exchange Commission (SEC) conducted their own investigations. There were also "shareholder and customer lawsuits." Not to mention article after article in the media detailing how poorly SCANA's upper management had behaved.

Soon, SCANA insiders were testifying against the management, and what they revealed was not pretty. It became clear that SCANA's statements to the Public Utilities Commission were just feel-good stories, meant to gloss over ongoing problems. The head of SCE&G's accounting team, who resigned after refusing to support the company's lies, characterized internal documents and emails as "creative writing" when speaking to the *Charlotte Business Journal*. Kevin Marsh, she revealed, had ignored her team's estimates of likely cost increases, preferring to put the best face on the project.

Marsh and other senior executives got most of the blame. The SEC, for example, accused senior executives at SCANA of perpetrating "a historic securities fraud" and of "repeatedly [deceiving] investors, regulators, and the public." V.C. Summer, the SEC declared, was "a tale of two projects." In the public version, "SCANA touted progress being made on the project" and these "false statements enabled SCANA to bolster its stock price, sell $1 billion in corporate bonds at favorable rates, and obtain regulatory approval to charge its customers more than $1 billion in increased rates to help finance the project." In the internal story, SCANA "knew that the project was significantly delayed, the construction schedule was unreliable and unachievable, and the company was unlikely to qualify for $1.4 billion in federal production tax credits because the new units would not be completed by the January 1, 2021, deadline for receiving the tax credits." Marsh and Byrne, the SEC insisted, "were at the center of this fraud."

The evidence was overwhelming, and both these executives had no choice but to plead guilty. A particularly crucial piece of evidence was a report produced by Bechtel (see chapter 5 for more about Bechtel's nuclear-weapons- and energy-related activities) that SCANA had commissioned. When the report concluded the project was "significantly off-schedule and over-budget," Marsh and his team suppressed it and never submitted it to the regulatory agencies, reported *Nuclear Intelligence Weekly*.

Stephen Byrne was the first to be held accountable, and in July 2020, he pleaded guilty to criminal conspiracy fraud charges. In November 2020, Marsh pleaded guilty as well and was eventually sentenced to two years in prison. Two Westinghouse officials, Carl Churchman and Jeffrey Benjamin, were also indicted in 2021.

The lesson for neighboring Georgia was to dig deeper into that hole. Its Public Service Commission voted to continue with the construction of the Vogtle reactors even after Westinghouse sought bankruptcy protection. The logic for pouring good money after bad was explained by Tim Echols, a member of the Georgia Public Service Commission and a strong supporter of the Vogtle project: "The moral of the [V.C. Summer project] story is that going over cost is bad, but canceling it is worse. That's the moral of the story, because all seven of the South Carolina commissioners have lost their job." Echols and his buddies preferred to keep their jobs to the disbenefit of Georgia's customers.

Such blatant disregard on the part of a commission that was supposed to look out for the public good can be described only as corruption. That was also the case in South Carolina, as well as in multiple other places around the world. Corruption, especially of the systemic kind, as opposed to individual malpractice, also helps us understand why some corporations continue to build expensive nuclear power plants.[3]

The beneficiaries of a failed project

In capitalist economies, in theory, an investor is rewarded with profits if the project is executed efficiently and should lose money if the project costs more than expected or is not producing value. The nuclear business works differently. Between 2009 and 2017, SCANA "paid over $2.5 billion in dividends to its investors" according to *Nuclear Intelligence Weekly*. And company executives, presumably for boosting the wealth of shareholders, received millions of dollars in compensation.

Executives also had what are called "golden parachutes" in corporate lingo—in other words, payments that would be automatically triggered in case of a sale of the company or a takeover. These payments were estimated at $28 million for Kevin Marsh. In other words, Kevin Marsh had plenty of money to buy fancy cars. And even contribute a little to charity organizations.

South Carolina is not an exception. In Georgia, too, Georgia Power and its parent Southern Company, have reaped billions in extra profit after they embarked on the wildly expensive Vogtle project. The grassroots environmental group Nuclear Watch South has been monitoring the financial reports of Georgia Power since 2008, and it documents an increase of annual profits by over 20 percent in the year 2011, when the charges for constructing the Vogtle reactors kicked into consumer bills. After that increase, profits have risen steadily from $1.1 billion in 2011 to around $1.8 billion in 2022.

It is little surprise, then, that these executives and shareholders continued to push for construction of nuclear plants, even after it had become painfully obvious that the projects would be grossly over-budget and greatly delayed in comparison to initial projections.

This has been the case with the whole push to restart nuclear construction that began with the 2005 Energy Policy Act under the Bush administration. As Peter Bradford, a former member of the US Nuclear Regulatory Commission, explained in January 2022: "The nuclear renaissance has not yet avoided a single molecule of carbon emissions … If the $40 billion spent on it so far had gone instead to the many less expensive and more reliable energy sources, the climate and the country would be far better off."[4]

Making money at the other end

Building power plants, or trying to do so, is not the only way to make money in the nuclear enterprise. As nuclear power plants age, they become prone to more problems and an increased risk of suffering accidents. Should the power plant's owners decide to do the right thing and shut down the reactor, they will likely embark on a series of processes called decommissioning, aimed at removing all the radioactive material from the reactor and the site to prepare the land to be used for other purposes. This geriatric stage in the nuclear reactor's existence is another cash cow for corporations.

Indeed, the accounting firm PwC (otherwise known as PricewaterhouseCoopers) has identified "nuclear decommissioning" as "one of the fastest growing segments of the nuclear power industry." As detailed in the *World Nuclear Industry Status Report 2021*, the growth will likely be most rapid in the United States, home to the most nuclear reactors—and the oldest ones.

As with all things nuclear, the expenditures involved in senior care for nuclear power plants are gigantic, and naturally, various corporations view decommissioning as a financial bonanza. In 2014, for example, Westinghouse officials—not that they have a lot of credibility—estimated annual revenues of a billion dollars from decommissioning. And a PwC analyst told *North Jersey Record* about the high "profitability potential" and highlighted the setting up of various hedge funds to invest in companies that deal with retired nuclear plants; some funds were worth hundreds of millions of dollars.

In 2019, a newly set-up company called Holtec Pilgrim LLC, a subsidiary of Holtec International, purchased the Pilgrim nuclear plant in Massachusetts from Entergy Corporation for a "nominal cash consideration"—corporate lingo for $1,000.[5] That might seem like a very low price for nuclear reactors that cost billions to build. But Pilgrim had been shut down in 2018 and was not going to produce any more electricity. Holtec would get no revenue and would have to spend money to safely manage the radioactive materials on the site. Seen in that light, $1,000 seems too much. Why would anyone pay anything? That is, literally, a billion-dollar question.

The answer has to do with the decommissioning fund. By law, nuclear power plant owners have to plan for decommissioning by setting aside money during the period that electricity and revenues are generated. These decommissioning funds can be sizeable; in the case of the Pilgrim nuclear power station, the fund has more than $1 billion, reported a 2018 article in the *Boston Globe*. When Holtec bought the reactor, it obtained access to this pot of money.

But it is not as simple as that. Historically, the process of decommissioning takes decades and cost estimates are often in excess of $1 billion. Thus, Holtec will be spending all the money that it can access. So, the question remains: Why would Holtec spend $1,000 to buy these old reactors?

Holtec has its fingers in many nuclear pies. In recent years, it has been promoting a small modular reactor design aggressively. Holtec's main product line, though, are special casks costing millions of dollars to store spent fuel. These casks are manufactured in the state of New Jersey, and Holtec was promised $260 million in tax breaks for locating its factory there.

Holtec plans to build a facility in the state of New Mexico to store casks containing radioactive spent fuel from decommissioned nuclear power plants. If the plan goes through, that is another source of profits for Holtec, because all the spent fuel from Pilgrim and the other nuclear plants that it has purchased would go there—packaged in Holtec casks, of course.[6]

Holtec also claims to be able to use various innovative technologies to quickly remove most of the materials on-site, allowing the facility to be fully decommissioned in a much shorter time frame. Six years or less after the reactor's final day at power is what the Holtec website promised in 2018.

The tight timeline means that workers will be forced to speed up and do their jobs faster, with possible repercussions to safety. According to an exposé published in December 2017 by the Better Government Association, this did happen at the former Zion nuclear power plant. Marilyn Lingle, a veteran nuclear worker, was working on decommissioning the plant. Lingle's job required her to measure radiation levels on scrap metal before it was recycled or sent to the landfill. By all accounts, she did her job meticulously, trying to ensure that the public was not endangered by radiation. But Zion Solutions wanted her to speed up; when she continued to prioritize scrupulousness over rapidity, the company first reprimanded her and then suspended her. Lingle filed a whistleblower complaint with the Nuclear Regulatory Commission, but her story came to a sad end: soon after the commission "opened an investigation, she was found dead in her truck in her garage ... [but the] coroner's office ruled that her death by carbon monoxide asphyxiation was an accident."

It is not just workers having to speed up. The quicker the timeline for decommissioning, the hotter the temperature and the higher the radioactivity levels of nuclear plant materials. In the more traditional approach to decommissioning, these materials would be allowed to stay on-site as they cooled, and some of the radioactivity would decay away. In the accelerated version, workers will have to handle more radioactive materials, increasing the radiation doses they would be exposed to.

Holtec's innovativeness is not limited to technical practices. The entity purchasing the reactors is a subsidiary of Holtec International and would have only limited liability. In other words, the subsidiary could declare bankruptcy and walk away from the job. This has raised questions about what might happen if the decommissioning fund is exhausted because of cost overruns and the site is not yet cleaned up to required standards.

So, the answer to the question of why Holtec would spend $1,000 to buy these old reactors has to do with the many ways in which it is going to profit from that purchase, while mitigating possibilities for losing money. Holding the bag in the latter case is, of course, the taxpayer.

Holtec has been promoting this business model and has entered into agreements with at least three utilities to take care of their reactors once these have been shut down. This model is evidently quite lucrative, as evidenced by Holtec being joined by SNC-Lavalin—a Canadian company notorious for having indulged in corrupt practices in Libya—in 2018 to form a US-based joint venture company called Comprehensive Decommissioning International. But that arrangement broke down a few years later.

Holtec's is not the only business model. Other companies have set up alternative versions. But in general the idea is to set up what scholars Marissa Bell and Allison Macfarlane describe as a "vertically integrated nuclear waste assemblage, in which a corporation (and its subsidiaries) own different parts of the nuclear waste supply chain." In their 2022 article in the journal *Energy Research and Social Science*, Bell and Macfarlane argue that the "purpose is to streamline activities for the most competitive advantage, and thus financially benefit themselves and their subsidiaries."

At the receiving end of this transfer of profit and risk are local communities in the locations where these materials are to be stored. In the United States, these include communities that have already experienced radioactive fallout from nuclear weapons tests conducted in the mid-twentieth century and Indigenous communities subject to nuclear colonialism.

There are also large amounts of money waiting to be made even further along, if and when a repository to store the waste is set up (see chapter 1). In 2019, the World Nuclear Waste Report reported that as of 2008 the total cost "of research, construction and operation" of the proposed geologic repository at Yucca Mountain over a hypothetical "150 year period—from when work started in 1983 through to the facility's expected closure and decommissioning in 2133" was estimated at $96.2 billion (in 2007 dollars), including the $13.5 billion already spent as of 2008.

In the United Kingdom, which is further along in the process of making more accurate estimates of its liabilities for decommissioning, the 2022 estimate ran to staggering £149 billion—over the next hundred years. Compare this with the 2018 estimate of £121 billion, and, better still, to the 2006 estimate of £51 billion, and you can see how fast these liabilities are growing.[7] Steve Thomas, a British energy analyst (see chapter 4), estimated the total bill to be as high as £260 billion.

Most of this cost is associated with cleaning up Sellafield, home to a reprocessing plant (see chapter 1). But this task is so complex that the top official at the UK's Energy Department told the Public Accounts Committee in 2015 that it was "impossible to know" the ultimate costs.

The vast majority of decommissioning costs will be paid for by taxpayers of the day stretching out over the next century and more. And then much of this money will be parceled out to private contractors. For these companies, then, nuclear waste is a profitable business opportunity, not a cost or a burden.

The climate-mitigation gap

How do these corporate maneuvers relate to the main focus of this book: nuclear power's potential role in mitigating climate change? Understanding the economic incentives motivating large electric utility companies—like SCANA or Dominion—will help us discern whether these incentives align with reducing the use of fossil fuels or not.

Despite the claims by nuclear advocates about the potential climate benefits of their favored technology, that factor played almost no role in SCANA's decision to build the V.C. Summer project. In his May 2008 application to the Public Service Commission, Kevin Marsh did mention how nuclear power would lower emissions, but his focus was on reducing the financial risk to the company should federal regulations impose a charge on emitting carbon dioxide. Marsh did not argue for expanding nuclear power capacity to produce less carbon dioxide—indeed, he could not, because SCANA was not planning to shut down any fossil fuel plants. He argued, instead, that some future government might force SCANA to do so, and building nuclear plants would help SCANA be seen as compliant.

In 2013, Marsh boasted that nuclear power "will allow us to achieve a very balanced generation portfolio. Once our two new nuclear units are complete, we anticipate our generation mix will be about 30% nuclear, 30% natural gas and 30% scrubbed coal, with the balance in hydro and some biomass." In other words, the SCANA leadership had no desire to get away from fossil fuels, nor any plan to increase renewables significantly.

SCANA was by no means unique in owning a mixture of fossil fuel power plants and nuclear reactors. Indeed, all the major nuclear-power-plant-owning utilities in the United States also own coal plants or natural gas plants or oil-fired plants. In 2022, Dominion, the parent company for SCANA, owned twenty power plants fueled by some fossil fuel. For Duke Energy, the owner of most of the nuclear plants in the Carolinas, plants burning natural gas or coal constituted 74 percent of Duke's electrical power capacity (as of December 2021) and produced 61 percent of the electrical energy generated in 2021.

For such businesses, nuclear energy is intricately tied with fossil fuels. As a result, utility companies tend to resist renewables, for it would mean giving up on decades of future profits.[8] Companies are particularly hostile toward distributed renewables, such as rooftop solar panels. Utilities have resorted to lobbying to defeat measures facilitating customers installing distributed renewables.

US utility companies are not alone. An academic study published in *Nature Energy* in 2020 showed that around the world these companies have "hindered the transition of the global electricity sector towards renewables, which has to date mostly relied on non-utility actors (such as independent power producers) for expanding the use of renewables." Electricity companies have also played a major role in promoting "climate denial, doubt, and delay" according to a 2022 paper in *Environmental Research Letters.*

Alliance building

In early 2009, Poland's government revived its Soviet-era nuclear power dreams, announcing that a first reactor should operate by 2020. Later that year, economy minister Waldemar Pawlak announced that a consortium led by the utility company Polska Grupa Energetyczna (the Polish Energy Group) would be set up by the end of 2010. Polska Grupa Energetyczna signed an agreement with Westinghouse—yes, the same company involved in the South Carolina fiasco—to explore constructing AP1000 reactors in the country. Nothing much came of that understanding, even though the US Department of Commerce put its weight behind Westinghouse, sending trade policy missions to the country.

Part of the problem was in Poland, where nuclear power was competing with vested interests in coal and prospectors for shale gas. In December 2021, Poland's nuclear planners selected two Baltic seaside villages as their preferred location. By then, the projected start date for the first reactor had been pushed back to 2033. But even that attracted Westinghouse, which had not won a single sales contract during the prior dozen years. Determined to maximize its odds, Westinghouse rushed to sign agreements with ten Polish companies. These companies include ZKS Ferrum, whose website describes it as "one of the largest Polish manufacturers of LPG tanks, vertical tanks and fuel tankers," and GP Baltic, which produces steel structures for wind turbines and ships.

In this renewed attempt to sell its reactors to Poland, Westinghouse is adopting the strategy of forming alliances with potentially useful partners. Its major competitor, GE Hitachi followed the same script in Canada. In October 2020, when Ontario Power Generation agreed to explore building a GE Hitachi–designed reactor in Canada, GE Hitachi promptly announced its separate agreements "with five Canadian companies: Aecon Nuclear, BWXT Canada Ltd., Hatch Ltd., Black & Veatch, and Overland Contracting Canada," according to *WilmingtonBiz*.

Why alliances? As argued, nuclear reactor vendors have one overwhelming goal: to persuade governments to force ratepayers and taxpayers to pay the high costs of nuclear reactors. Having multiple groups lobby for this goal makes the job easier. Alliances also helps nuclear advocates resist policies favoring renewable energy. Having multiple, seemingly independent entities put out propaganda in support of nuclear power helps persuade the public that the expensive and risky source of electricity might be necessary, or even a good idea.

Companies tend to join hands especially during periods when the nuclear reactor market is floundering. In 1987, as the industry reeled from the simultaneous impact of the Chernobyl catastrophe and over-capacity in reactor building, Martin Spence explained in *Capital and Class* that nuclear companies formed "defensive alliances" at "both national and international levels" as they bided their "time and [waited] for an upturn." Companies also gang up during the early stages of nuclear programs in a new country, as in the case of Poland. And in South Korea, where American nuclear companies and Korean business conglomerates (*chaebol*) joined hands due to a confluence of financial interests.

Interests of nuclear commerce trump even defense partnerships. In August 2022, Korea Hydro and Nuclear Power and Russia's Atomstroyexport entered into a $2.25 billion contract. Korea Hydro and Nuclear Power was to supply materials and construct the turbine building at El Dabaa, Egypt's first nuclear plant. Atomstroyexport had just begun building the El Dabaa plant in July 2022, five years after the contract—this one for $29.4 billion—was signed between Russia and Egypt.

All of this was amid the concerted US effort to sanction Russia over its attack on Ukraine.

Companies often target universities. To smooth its entry into Indonesia, ThorCon, which is peddling a thorium-based nuclear plant built on a ship, signed agreements with the renowned Bandung Institute of Technology and a number of other universities. ThorCon's other partner is even more powerful: Indonesia's Defense Ministry.

Alliances also feature in countries like China, with an entirely different market system. China has two large "state-owned enterprises" that compete to build reactors. The older one, China National Nuclear Corporation (CNNC), has historically participated in both civilian and military nuclear activities. The second, China General Nuclear (CGN), was originally created to operate China's first imported nuclear power plant at Daya Bay. Despite their fierce rivalry, both companies came together to design the Hualong-One reactor, and formed an equally owned joint venture, Hualong Corporation, to market this design abroad. Increasingly, the two companies swap senior officials.

In India, too, only state-owned entities can operate nuclear plants. The primary entity, Nuclear Power Corporation, has close ties with major private companies like Larsen & Toubro, Walchandnagar, Godrej & Boyce, and Tata Consulting Engineers. The public support that these companies offer bolsters the government's ideological clout. For example, in May 2017, when India's cabinet, chaired by Prime Minister Narendra Modi, announced plans to construct ten nuclear reactors, the director of Larsen & Toubro called the move "bold and historic," while the chief operating officer of Godrej & Boyce, termed it a "visionary" step. Godrej & Boyce's praise may also be motivated by a vision of future profits: they got a contract for ₹4.7 billion in 2021.

The most important of such alliances, if one is evaluating whether nuclear power will solve climate change, is with the fossil fuel industry. We have already discussed how electric utility companies in the United States own both nuclear plants and fossil fuel plants. The rest of the nuclear enterprise, too, gangs up with fossil fuel companies.

In June 2020, as the oil and gas industry reeled from the economic and social impacts of the COVID-19 pandemic, the Nuclear Economics Consulting Group published a report advising "oil and gas majors" to invest in new nuclear plants, specifically small modular and advanced reactors. The report asked a series of rhetorical questions—for example, "What group is under a withering attack by climate change activists?" —all of which had the answer: "oil and gas majors." It then went on to advise these companies to invest in nuclear power to dispel "their perceived (and sometimes actual) climate change insouciance."

Naturally, these companies—even ones that talk up nuclear power as the answer to climate change—are unwilling to call for strictly regulating carbon dioxide. And it is not just carbon dioxide. In April 2022, the Utilities Solid Waste Activities Group, comprising electric utilities with fossil fuel assets, sued the Biden administration over its attempts to stop leakage from ponds holding ash from coal plants. The ash has been seeping from these ponds and contaminating water supplies, but corporations are unwilling to add a lining to these ponds to prevent this contamination. More generally, the group would like to get rid of requirements for the safe disposal of ash from coal-fired power plants. Several of the members of USWAG are nuclear plant owners, including Southern, Santee Cooper, Tennessee Valley Authority, FirstEnergy, Dominion, Entergy, and Exelon.

While many of these companies profess great concern about climate change, especially when lobbying for subsidies to nuclear power plants, it is in their economic interests to maintain fossil fuel use and keep the current energy system more or less intact.

The attractions of nuclear energy

Coming back to the question I started with, about why companies get involved in nuclear energy despite bleak economics, what I have argued is that companies view nuclear power as an attractive investment as long as the exorbitant costs of building and operating nuclear plants are foisted on the pocketbooks of the public, either in the form of higher electricity bills or in the form of taxes, and the multiple environmental and economic risks associated with the technology are socialized. The profits, of course, accrue to companies.

Underlying many decisions concerning nuclear power, then, is a simple fact: many powerful organizations and some governments benefit economically and politically from the nuclear energy business. This core reality explains why these organizations and their allies support generating power from the atom.

It is not just electricity companies that profit from nuclear power. The list of other corporate interests vested in nuclear power is long: starting from financial ones buying up and selling businesses, to consulting companies offering advice about how to go about building a nuclear plant, to Wall Street banks loaning money, to law firms writing contracts, to insurance companies that take a cut for agreeing to hold part of the bag in the event of an accident, to ... I could go on in this fashion.[9]

The Canadian company Brookfield Business Partners entered this nuclear investment club only in 2018, when it purchased Westinghouse. The acquisition occurred soon after Westinghouse had just received bankruptcy protection following the debacle in South Carolina. Some years later, in May 2022, Brookfield put up Westinghouse Corporation for sale. *Nuclear Intelligence Weekly* reported the logic put forward by Brookfield's CEO to financial analysts: "Our job is sort of done here, and we're on to the next five great investments where we think we can replicate the same type of outcome. To do that, though, we need to recycle capital, generate capital. That's our business."

One wonders whether the CEO had just brushed up on *Capital*, for his logic sounds very much like how Karl Marx described the thought processes animating capitalists in his magnum opus. What differentiates Brookfield is the source of profits (or delta-M, to be precise), which are derived not from exploiting workers and manufacturing commodities but from growing hype (and possibly by exploiting workers in PR and media firms). In the case of Westinghouse, the hype from Brookfield's executives involved half truths. Two of these, about Westinghouse being "the only alternative to Russian companies to supply fuel to Russian reactors outside Russia" and about how Westinghouse might export AP1000 reactors to Poland or the Czech Republic, explain the timing—Brookfield announced the sale just two months after Russia's attack on Ukraine.

Reactors, the central element of the nuclear landscape, are hugely expensive. The high cost is welcomed by corporations, for it offers the possibility of large contracts and massive profits. Indeed, much of the money is made well before the facility is completed and before the first unit of electricity flows out of the plant—much in the manner highlighted by Anthony Trollope in his novel about the robber barons of the nineteenth century mentioned in the epigraph.

Nowhere has this benefit of investing in nuclear power been so apparent as in the case of South Carolina discussed earlier. SCE&G's persistent, even stubborn, support for building the two AP1000 reactors is best explained by the company's executives and shareholders making handsome amounts of money, at the expense of ratepayers. Nothing else can so well illuminate why the corporate suite of SCE&G ignored all the warning signs.

Some readers might wonder whether one could attribute what happened to a lack of awareness about the problems in

China's corresponding reactors, or just poor project management. Perhaps simple due diligence might have caught these issues and planned for them. But that diagnosis is missing a key feature of the nuclear landscape. V.C. Summer was a flagship project of the so-called nuclear renaissance, and everyone involved in that enterprise wanted the reactors to start operating in the shortest possible time span between when the plan to build nuclear reactors was announced with much fanfare and when they came online to even greater jubilation. There was simply not enough time for even completing the first round of projects, let alone examining them for possible lessons.

The same constraint, of trying to build as soon as possible, will also come in the way of trying to get any of the proposed small modular reactor projects built as carefully and thoughtfully as possible. As described in chapter 6, these reactor designs constitute the next round of nuclear hype, and once again proponents have allotted no time whatsoever for learning from projects. Again, nuclear advocates cannot be measured and careful in their planning; for SMRs to be portrayed as solving climate change, proponents have to necessarily argue for building them by the hundreds and thousands within the next couple of decades. Even as paper projects, it is obvious that this buildout will not proceed smoothly. But nuclear advocates cannot acknowledge this reality. To do so would automatically eliminate small modular reactors from realistic plans for lowering emissions at any meaningful scale or time frame.

Whether reactors are small or large, the key question for nuclear investors is how to make the public pay. This task is made easier when the nuclear industry allies with other politically powerful industries. These alliances are predicated on a simple principle: the whole can exert more political clout than the sum of the parts. Alliances allow companies and sectors to build off each other's strengths and present a unified front in the face of public challenges to their legitimacy.

The alliance that most affects action on climate change is the one between the nuclear industry and the fossil fuel industry. This alliance is rooted in the financial overlaps between corporations that invest in and profit from both these sectors. In many cases, there is no need for alliances, since it is the same corporation that is profiting from fossil fuels and profiting from nuclear power. What also binds the nuclear and fossil fuel sectors is a common opposition to renewable energy, especially when this involves newer forms of decentralized generation, or any major shift in the electricity supply model. The most important alliance partner for the nuclear industry, however, is the government, and we turn to that actor in the next chapter.

4

Enabling Moneymaking, Singing Praise: Governments and Nuclear Power

We are on a pathway to global warming of more than double the 1.5-degree limit agreed in Paris ... Some government and business leaders are saying one thing—but doing another. Simply put, they are lying ... And the results will be catastrophic.

UN Secretary-General António Guterres, April 2022

[The] neoliberal state facilitates a vast and ongoing redistribution of resources from public to private coffers. The idea is simple: wealth and power are accumulated by a few in the act of dispossessing—that is, stealing from—the population at large. Of course, our status-quo stories aren't wont to call it "stealing," given the fact that it's perfectly legal.

Julie Wilson, *Neoliberalism*

"Boris Johnson Goes Nuclear with Swansong Energy Investment," read the Reuters headline. With less than two weeks to go before his term as lame-duck prime minister ended, Johnson offered £700 million to build two new reactors known as Sizewell C. To be constructed by France's state-owned utility, Électricité de France (EDF), Sizewell C was estimated to cost more than £25 billion. Johnson had earlier contributed £100 million of public funding for the nuclear plant. As well, he had signaled his willingness to get the project going in myriad other ways.

Johnson was not the first British prime minister in recent years to put the UK government's financial and policy shoulders to the nuclear wheel. Sizewell C and every other nuclear project being considered in the United Kingdom are indebted to Tony Blair. The incentives he put in place for nuclear power resulted in plans being developed for as many as ten nuclear reactors, and the start of construction of the Hinkley Point C nuclear plant, the UK's first reactors since 1988.

But between 2013, when the UK government and EDF settled the terms of Hinkley Point, and Johnson's 2022 announcement, not a single new project took off. Despite the UK government making its interest in nuclear power abundantly clear, no company embarked on building reactors absent substantial funding. Examining this history more closely illuminates how the fortunes of nuclear power critically depend on government support, the focus of this chapter.

Governments promote the interests of nuclear energy producers by providing subsidies of different kinds, and skewing electricity markets. Such financial and policy backing undergirds nuclear power, both in countries where the electricity business is privatized (e.g., the United States, the United Kingdom) and—somewhat self-evidently—in countries where it is dominated by the government (e.g., India, China).

Governments also support nuclear power by disseminating propaganda: offering justifications for why nuclear energy is necessary, including by projecting impending shortages of energy; making claims about why nuclear power should be attractive to different groups in society—for example, by touting how many jobs it creates; and by positing nuclear energy as a marker of modernity and status, especially in developing countries. Both forms of support tend to be offered by all mainstream parties, ranging from the technocratic liberal end of the spectrum (Blair or US Presidents Obama and Biden) to the right wing, known for their antipathy to addressing the climate crisis (Johnson or US President Trump).

To be sure, not every government supports nuclear energy in this fashion. Indeed, the vast majority of governments around the world don't do any of this. But those countries also don't deploy any nuclear power plants. Most don't even aspire to building reactors in the future. But countries that have never considered nuclear power don't feature much in discussions about the technology. Hence the focus here is on countries with operating nuclear plants or those that seek to build reactors.

I begin this chapter by using the example of the Hinkley Point C project in the United Kingdom to illustrate nuclear power's dependence on government action. That project would not have proceeded if it had not been subsidized by taxpayers, if consumers had not been forced to pay higher electricity tariffs to procure nuclear power instead of cheaper alternatives, and if rich investors had not been enticed to put in billions of pounds—all measures only the British Parliament and politicians could deliver. Nuclear energy wouldn't survive without all these forms of political patronage, neither in the United Kingdom nor in other countries. I explore some reasons for why governments are so keen on nuclear power elsewhere in this book.

Setting the ball rolling

Tony Blair's rise to political power started at the 1994 Labour Party conference, where he announced his intention to amend Clause IV of Labour's constitution.[1] Introduced as part of the 1918 document that declared the aims and values of the party, this clause committed Labour to public ownership—and was seen widely as the party's commitment to socialism. By contrast, Blair committed Labour to "the enterprise of the market and the rigour of competition," as a 2020 article in *Jacobin* explained. In the case of nuclear power, however, this was just rhetoric, as Blair's efforts to prop up the industry were to show.

Blair was not always so committed to market competition. Five years earlier, as Labour's energy spokesperson, he had attacked the privatization of nuclear power in the United

Kingdom under Prime Minister Margaret Thatcher's regime. It will be "the most expensive mistake in the history of privatization," the *Guardian* reported him declaring. The odd thing about that attack was that Thatcher was fond of nuclear power, having weaponized atomic reactors as part of crushing the protracted strike by coal miners earlier that decade.[2] What went badly for nuclear power, though, was Thatcher's espoused preference for the free market. The market won. Not a single nuclear reactor was built in the UK after privatization.

Nor was Blair always in favor of constructing nuclear plants. In 1987, he attacked the Conservatives for their plans to construct more reactors, pointing out that "radioactive waste is itself a major environmental problem. And one for which we have no easy answer at present." During the 1997 elections, the Labour platform stated: "We see no economic case for the building of any new nuclear power stations."

Blair changed his position once he came to power, ordering a unit he had set up in his own office to carry out a review of energy policy. Why did he change? One really doesn't know, but British journalist Jonathan Leakey has argued that this shift was a result of Blair appointing Sir David King, an evangelist for nuclear power, as his chief scientific advisor. Once Blair had flipped, though, he came up with a number of rationalizations for his advocacy of nuclear power, mostly focusing on the need to lower emissions and to reduce imports of fuels—what is glibly referred to as "energy security."

The policy reversal was remarkably rapid, carried out undemocratically over a space of a few years and in the face of clear evidence of the problems with the proposed strategy. In 2001, the *Independent* reported that an early draft of the review Blair commissioned had concluded that "nuclear power seems likely to remain more expensive than fossil-fuelled generation" and that "nowhere in the world have new nuclear power stations yet been financed within a liberalised electricity market." Blair insisted on nuclear power being included in future energy plans and keeping open the option of building new nuclear reactors. When the report was finally published in 2003, it acknowledged the unattractive economics and the unresolved problem of nuclear waste but committed to keeping the option open.

But the timing wasn't good. In 2001 and 2002, the nation's two leading nuclear enterprises, British Energy and British Nuclear Fuels Limited (BNFL), underwent spectacular financial collapses, largely a result of nuclear power being subject to market scrutiny. BNFL, for example, was unable to meet its liabilities, an estimated £34 billion, for cleaning up the radioactive messes at the many facilities it owned. British Energy issued an insolvency alert after its stock price fell from a high of £7.30 in 1999 to a low of £0.54 in 2002.

And on September 24, 2002, Blair told the UK Parliament that Iraq was acquiring nuclear weapons and other weapons of mass destruction, thereby committing his country to the Bush administration's disastrous attack on Iraq. Those false claims were the focus of media and political attention over the next couple of years, leaving little time for other major initiatives.

Once those storms had died down, in November 2005, Blair went back to the business of promoting nuclear energy, starting with a speech to the UK's wealthy, the Confederation of British Industry. There, Blair characterized the role of government as setting "the climate for business," and investing "taxpayers' money to create the right human and physical capital." He then went on to announcing yet another review of energy policy, only two years after the earlier one concluded about the inadvisability of new nuclear construction. But this one, Blair signaled, would be different, because it was tasked to look into whether the government should "facilitate the development of a new generation of nuclear power stations."

By July 2006, the evidently fast-paced review delivered what Blair promised. It concluded that "new nuclear power stations would make a significant contribution to meeting our energy policy goals." Although there was no breakthrough in nuclear construction in the years since the earlier document, this new review claimed that the economics of nuclear power "looked more positive." Fueled by this hype, the report simply assumed that the capital cost of building a reactor would be between £850 and £1,600 per kilowatt, and the cost of electricity from these would be in the range of £30 to £44 per megawatt-hour.

Construction costs were made more explicit in a white paper published in 2008, which assumed that a single 1.6 GW reactor would cost somewhere between £2.0 billion and £3.6 billion to build. It went on to warn about the "risk of cost over-runs in construction" but averred that its analysis was "based on conservative assumptions." The white paper also stated that it was up to private companies to decide whether they wanted to invest in nuclear reactors.

As it transpired, the cost assumptions in the white paper turned out to be anything but conservative. Nevertheless, it was good enough to allow the UK to embark on building new nuclear plants. Once the process started, it acquired a momentum all its own. As Jonathan Swift put it three centuries ago, "Falsehood flies, and the Truth comes limping after it; so that when Men come to be undeceiv'd, it is too late; the Jest is over, and the Tale has had its Effect."

Prime Minister Gordon Brown, Tony Blair's successor, helped the process along by publicly calling for eight new nuclear plants within the next fifteen years—that is, by 2023. An anonymous Downing Street official explained the subtext to the *Guardian*: "The industry will not make the long-term investment required to build a new nuclear power station if they think the government is not totally committed to nuclear energy." In other words, the government wanted to reassure those with capital and offer handsome rewards if they invested in nuclear power.

Nor did the UK government leave the investment decision as a matter for private corporations. By 2010, Ed Miliband, the energy and climate change secretary, was calling for "a more interventionist energy policy." The interventions included changes to the planning system, regulatory process, and the emissions trading system, all meant to enable easier business for nuclear companies. For example, a new Infrastructure Planning Commission was set up to "fast track" applications, and in 2009, two proposed nuclear power plants, one each at Hinkley Point and Sizewell, were chosen among the first round of projects.

At a press conference soon after, Steve Thomas, then a professor at the University of Greenwich Business School and a member of its Public Services International Research Unit, highlighted the financial risks involved. Steve is one of the most perceptive analysts of the UK's nuclear and energy policies, and someone with whom I have collaborated. Over the years, I have relied on his assessment of nuclear power policies and energy policies more generally, both in the UK and around the world.

At the press conference, Steve posed the important question: "If things go wrong, who will pay?"[3] The answer should come as no surprise: taxpayers. When it went bankrupt, Steve pointed out, British Energy was rescued at a cost to future taxpayers of more than £10 billion. The public has also been presented with a "bill of £80bn and rising" for decommissioning nuclear plants. (As mentioned in chapter 3, that bill has gone up significantly since then, and continues to rise.) Things started going wrong for taxpayers very soon.

The attitude toward nuclear energy and big business did not change when the new government, a coalition between the Conservatives and Liberal Democrats, took over later that year. In October 2010, Minister of Energy Chris Huhne promised to "create the right environment for business to invest in the energy market." The similarity of the rhetoric with Labour was not accidental. In recent decades, regardless of which party was in power, the UK government has been ready to roll out the red carpet for wealthy entities willing to put money into building nuclear power stations.

Huhne had to pretend on another account. Although he presumably knew that the government's nuclear power plans would never move forward without subsidies, he asserted that "there will be no public subsidy for new nuclear power." His government was imposing harsh austerity measures for the larger population of the United Kingdom; it would simply not seem fair if the government was seen to be doling out money to energy companies.

At least at that time. Huhne's party and the earlier Labour government had so far promised only to ease the process of construction. It took until 2015 for the government to admit, using a double-negative clause, that it was "not continuing the 'no public subsidy policy' of the previous administration."

At the receiving end of those potential payments was Électricité de France. EDF had purchased British Energy in 2008 for £12.4 billion. As part of the deal, EDF agreed to meet the costs associated with nuclear waste management and decommissioning for any new plants it builds in Britain but refused to take on most of the costs of waste and decommissioning work for existing reactors. Those costs are to be borne by a company called the Nuclear Liabilities Fund, which is mostly financed by the UK government—that is, British taxpayers—with some marginal contributions from EDF. These were the costs that Steve Thomas was referring to.

Propping up Hinkley Point

The route that the Conservative government chose to subsidize Hinkley Point and other nuclear plants was through what it termed "Contracts for Difference." The idea was to pay EDF a fixed price for each unit of electrical energy for the next thirty-five years. But this fixed price, Steve Thomas explained in a 2016 paper in *Energy Policy*, would come partly "from the market" and partly from "a consumer subsidy," with the proportion depending on what the market price of electricity was on any given day.

How much was this fixed price to be? This was the subject of what the *New York Times* described as "months of dickering between the British government and EDF" in a March 2013 article. At the heart of that dickering was the profit rate. The EDF desired a 10 percent return on its investment, and the British government wanted to offer only 8 percent. Settling that dispute was a litmus test for which party was more attached to the project.

Early in 2013, Vincent de Rivaz, the head of EDF's British division, went down to Downing Street, the *New York Times* reported.[4] In a meeting with the cabinet secretary, he threatened to walk out of the project unless the government met EDF's demands. The ultimatum worked; the government was the first to blink. It offered a fixed price of £92.50 per megawatt-hour (at 2012 prices), slightly lower than the £100 per megawatt-hour that EDF was seeking, but more than twice the upper end of the £30 to £44 per megawatt-hour that the report commissioned by Blair predicted. The figure was linked to inflation, which means that it amounts to £127.11 in September 2023 prices, using the inflation rates recorded by the Bank of England. EDF was guaranteed an inflation-linked price for thirty-five years.

The £92.50 figure was also just over twice the average wholesale electricity prices in the UK, which was £46 per megawatt-hour in 2012. In other words, had Hinkley Point been generating power that year, consumers would have had to fork out the remaining £46.50 for each megawatt-hour of electricity the plant would have fed into the grid.

Being able to extract such a promise is a measure of the extent to which corporations like EDF can take advantage of governments, especially if these have strongly expressed commitments to nuclear technology. In this case, the UK government had already signaled that it badly wanted the project to move forward. Just the previous year, the chair of the government's Energy and Climate Change Select Committee had publicly called upon the prime minister to clearly confirm "that nuclear is a part of the policy and the Government is going to do what's necessary to incentivise it."

The underlying reason for the cost of electricity going from the predicted £30 to £40 per megawatt-hour to £92.50 per megawatt-hour was the escalation in the estimated cost of constructing the Hinkley Point nuclear plant. As I described in chapter 2, this phenomenon is all too common with nuclear construction projects. In the case of Hinkley Point, the cost more than doubled from the theoretical value of £3.6 billion per reactor assumed in the energy policy review to £8 billion in 2013.

And it kept increasing. Meanwhile, EDF's balance sheets were not looking so good. The company unsuccessfully tried to enter the US nuclear power plant market, with plans to build a plant in Calvert Cliffs, Maryland, that came to naught. Speaking to *Bloomberg Business* in 2013, its CFO admitted to a loss of €2 billion between 2009 and 2012 because of the escapade. EDF shares had plummeted, losing 85 percent of their value between 2007 and January 2013, with the rating agency Moody's downgrading EDF's perspective from "stable" to "negative" in December 2012.[5]

The governments of the United Kingdom and France, indirectly through its ownership and financing of EDF, had done all they could within the limits of political decorum. It was time for another government to help with facilitating Hinkley Point.

In 2015, the United Kingdom did a "Nixon goes to China" act; Chancellor George Osborne went to Beijing seeking to set up the UK as "China's best partner in the West." There he offered a £2 billion guarantee toward Hinkley Point to entice the Chinese government to invest in the nuclear power station.

The following month, China took a one-third stake in Hinkley Point. The decision to involve China was, and continues to be, contested based on geopolitical grounds and on concerns that the Chinese may be able to interfere with the British grid. But it was a political quid pro quo. The UK government was desperate to get construction of Hinkley Point going. Its primary constraint was that EDF simply did not have the capability to finance this. China, on the other hand, had plenty of capital looking for markets.

China also had an underlying reason: a desire to enter nuclear export markets. This motivation had intensified with the increasing capacity of the Chinese nuclear industry and the slump in nuclear markets around the world that followed the Fukushima disaster. As a senior official put it in 2013, "History has given China an opportunity to overtake the world's nuclear energy and nuclear technology powers."

But there were hurdles to be overcome before that opportunity could be seized. Till now, China has exported nuclear reactors only to Pakistan.[6]

When it came to other countries, Chinese companies faced two main challenges: one is the widespread perception that Chinese-designed reactors were not as safe as Western reactors.[7] The other challenge is the concern that Chinese companies might be able to build reactors in China but have no experience outside the country. It is to address the second concern that Chinese companies were using their financial clout to become investors in power plants. As a Chinese academic explained it to *Xinhua*, China's official press agency, "Success in the British market will set a good example for ... future exploration of other foreign markets, like Southeast Asia, the Middle East and Africa."

Thus, it was that three governments came to support the only nuclear power plant being built in the United Kingdom.

Channeling public money to private companies

What are the lessons of this history of Hinkley Point, and how does it apply to nuclear plants elsewhere? I start with how governments offer financial support to nuclear power.

In the case of Hinkley Point, the government enabled the finances that made the project possible. Without the guarantee of a fixed price for multiple decades, EDF would not invest billions of pounds. In turn, the government passed on those costs to future consumers of electricity, who will be paying more for their power than they would have otherwise. Those "top-up payments," a report from the House of Commons Committee of Public Accounts estimated, will "cost consumers around £30 billion over the 35-year contract." The International Institute for Sustainable Development estimated in 2016 that if loan guarantees and decommissioning costs are taken into account, EDF could be receiving as much as £58 billion in subsidies.

Even this wasn't good enough for EDF and other nuclear vendors, when it came to building more reactors. Despite much talk about nuclear plants at the Sizewell, Bradwell, Wylfa, and Moorside sites, not one company moved forward to building any reactors.

So, in 2019, the UK government started considering a new model for forking over public money to nuclear corporations: the regulated asset base (or RAB) model. The idea was similar to what happened in regulated states such as Georgia and South Carolina in the United States (see chapter 3). Consumers would pay upfront for the reactors, well before they started generating electricity, including for overruns or construction delays. They would "also compensate nuclear investors if the project were scrapped," reported the *Observer*. For its part, EDF explicitly admitted the RAB mechanism would "make the project more attractive" to investors.

The flow of profit is what sets apart the investors from the consumers. Consumers are not treated like financial firms or investment banks. Instead, under the RAB model, consumers provide the financing for projects "at zero interest," bearing "some of the risk associated with construction costs," but without being "paid to hold these risks in the way investors would be." In June 2019, when news of this scheme broke out in the media, Beyond Nuclear's Linda Pentz Gunter aptly quipped on Facebook: "It's called RAB but it should actually be called ROB."

When it comes to financially supporting nuclear power companies, the United Kingdom is no exception. The French government has routinely propped up EDF, including with an injection of €2.2 billion in 2022.

In the United States, nuclear companies benefit from subsidies, grants, and bailouts from state governments and the federal government. Again, 2022 was a particularly lucrative year. In April, the Biden administration offered $6 billion through its Infrastructure Investment and Jobs Act. Three months later, the Inflation Reduction Act included a "zero-emission nuclear power production credit" that offers up to $30 billion to nuclear utilities according to the Congressional Budget Office. The Nuclear Information and Resource Service offers a higher estimate: $53.5 billion through 2032; what is more, these "taxpayer dollars would accrue to a very small number of large power corporations and utility holding companies. Over 85% of the total would be claimed by 12 companies ... with $20.0 billion ... by one corporation, Constellation, which owns 21 merchant reactors."

States did their bit to ensure profits to utility companies with nuclear plants. In August 2022, California's governor, Gavin Newsom, offered a "$1.4 billion forgivable loan" to Pacific Gas & Electric to keep the Diablo Canyon nuclear plant open.[8] This was in addition to $75 million the previous month. In November 2021, Illinois offered $694 million to Exelon to keep reactors operating. All of this is on top of earlier bailouts to the tune of roughly $14 billion in the last decade from New

York ($7.6 billion), Illinois ($2.4 billion), New Jersey ($2.7 billion), and Connecticut ($1.6 billion).

Bailouts by state governments in the United States were so profitable that even anti-government interventionists like the Koch brothers invested in firms that operated nuclear plants.[9] The rationale for these wealth transfers to companies was explained in June 2019 by then Secretary of Energy Rick Perry in the context of a debate over subsidizing the Davis-Besse nuclear plant in Ohio. As reported in *Utility Dive*, Perry called upon states to craft tax and regulatory policy to send "the message that capital is welcome in your state." In other words, all the talk about avoiding carbon dioxide emissions is really about making sure that capitalists are kept happy.

The beneficiaries of these subsidies are all giant corporations. Pretty much every nuclear plant in the United States is owned by companies that have market capitalizations of tens of billions of dollars.[10] Even though I have used the common term "subsidies," a more appropriate term might be "corporate welfare."[11] These giant corporations and various associated organizations have engaged in extensive lobbying and propaganda campaigns to get governments to pass legislations that make consumers pay more for the electricity they use.[12] In turn, the interventions by governments have increased the financial might of these corporations, which in turn contributes to their clout in state and national policymaking and their ability to fund their advocacy efforts, and even to pay politicians tidy sums of money.

Promoting exports

Nuclear vendors have always competed fiercely for reactor orders. Competition has become more intense as prospects for nuclear power fade in the United States and much of Europe. Supplier companies have therefore been scouring regions of the world that are seen as more promising. China and the Middle East are leading examples.[13]

Indicative of the high stakes involved in nuclear reactor deals is the involvement of very senior government officials, all the way up to heads of state, at various points in the negotiation. As the *New York Times* reported in 2004:

> In recent months, a procession of political leaders has pressed China to favor power plant designs and equipment from their home countries. They have included President Jacques Chirac of France; former Prime Minister Jean Chretien of Canada; Viktor Khristenko, who was named fuel and energy minister in Russia on Tuesday; and dozens of less-prominent officials. President Bush even raised the virtues of American nuclear technology with the Chinese prime minister, Wen Jiabao.

A few years later, South Korean president Lee Myung-bak visited the United Arab Emirates to successfully advocate for the Barakah nuclear plant contract. The deal "couldn't have been achieved without President Lee's active salesmanship and the strong support of the government," the CEO of Korea Electric Power Corporation later acknowledged.

Thanks to WikiLeaks publishing a cable from the US embassy in Beijing, we know that officials pushed the government to do more: "Effective advocacy for US nuclear suppliers is essential to ensuring access to China's rapidly growing civil nuclear power market ... Regardless of how the United States decides to advocate, it should be done continuously and from a high level in order to keep up with the French and Russians."

Not that government officials needed the prodding. Many of them are proactive, utilizing multiple units of the government to further the interests of the nuclear sector. Consider this speech delivered by Rose Gottemoeller, the acting undersecretary for arms control and international security in the Obama administration, to the Nuclear Energy Institute:

It may not be the first impulse of export firm executives to think of the U.S. Government as a business asset, but there is much that we can do to help. We are developing what we call a "Team USA" approach to civil nuclear engagement abroad …

Another service that the government can provide is advocacy. Once a potential nuclear project is approved for advocacy by the Department of Commerce's Advocacy Center, the State Department and other U.S. government agencies can, through active diplomacy with the host country, put U.S. Government support behind the American bidder. Even when more than one American firm is bidding on a nuclear power plant, we may be able to engage in generic advocacy, expressing to the host government our support for a U.S. firm winning the contract. We also try to ensure that a foreign government's decisions are being made in a transparent manner on a "level playing field." Our diplomatic posts are sensitive to any evidence that undue influence is affecting a host government's decision, and those posts are prepared to protest unwarranted discrimination against U.S. sellers.

There are a number of other steps that the Administration has taken to ensure that our nuclear exports receive the attention they deserve. The Department of Commerce has established a Civil Nuclear Trade Initiative, the goal of which is to identify the U.S. nuclear industry's trade policy challenges and commercial opportunities and coordinate public-private sector responses to support the growth of the U.S. civil nuclear industry.[14]

Team USA continued far beyond President Obama's tenure. During the Trump presidency, government officials pushed hard on Eastern European states to purchase US nuclear reactors, signing agreements with Bulgaria and Romania in October 2020. The Biden administration followed up with a $14 million grant to Romania to encourage it to embark on building (US) small modular reactors.

Governments have used financing arrangements, often through Export Credit Agencies, to make it easier for importing countries. Canada's Export Development Corporation, for example, has loaned money to India, Pakistan, Argentina, Romania, South Korea, and China as part of its strategy to promote CANDU reactors. The United States provided such financing for fifty of the sixty-three export orders for US nuclear reactors between 1955 and 1980.

More recently, the Trump administration changed the rules governing the Development Finance Corporation to allow it to fund nuclear projects. It is "the only government development agency in the world" to do so, suggests the *Hill*. Such institutions are supposed to be focused on improvements in the world's poorest countries, none of which are likely to benefit from expensive nuclear power plants. But the proposal allows the nuclear industry to profit at the expense of the taxpayer while pretending to alleviate poverty. And of course, this change was touted as "crucial to meeting climate and energy leadership goals," ironically by two senators, Lisa Murkowski and Joe Manchin, well known for their support for fossil fuel industries.

Russia has taken this financing approach to new heights. In Bangladesh, it loaned 90 percent of the cost of two VVER-1200 reactors, and this loan is to be paid back over "the next 28 years with an 8-year grace period," clearly an investment for the long term. Rosatom is paying the full cost of constructing and operating four VVER reactors in Turkey, hoping to make its money by selling electricity—the build-own-operate model.

Rosatom has also benefitted from forceful and persistent governmental action to secure contracts. As *Nuclear Intelligence Weekly* explained in 2015: "Rosatom has managed to muscle out competitors in tentative newbuild markets from Bangladesh to Algeria through the use of the government pen: in each case it has pushed through a series of bilateral agreements, with each one more detailed than the previous ... pushing for enough intergovernmental deals that a commercial contract is ultimately inevitable."

As a result, Russia has cornered the nuclear reactor export market. A year after the Fukushima disaster started, the head of Russia's nuclear supply company, Rosatom, announced having foreign orders worth $50 billion. By September 2015, that had increased six-fold. The Kremlin website recorded Rosatom's CEO crowing about "orders for 30 nuclear power plant units in 12 different countries ... worth a total value of more than $300 billion" to President Vladimir Putin.

Not all of these reactors will be built and one has to discount for self-serving advertisement. Nevertheless, these statements testify to the power of cheap financing and political support from governments in boosting nuclear hype.

In countries that are importing these reactors, the availability of "easy money" leads to politically opportunistic elites pushing for these nuclear projects. Such elites range from government officials and politicians to corporate conglomerates to even media houses that are seeking to further globalization and trade. Such drivers motivate not just nuclear power projects but a variety of megaprojects. Political scientists Jessica Liao and Saori Katada, for example, detail similar dynamics in the case of high-speed rail projects funded by China and Japan in Southeast Asia.

Governments interested in acquiring nuclear plants cultivate such elites, enticing them with ways of profiting from these projects. In a 2014 article in *Monthly Review*, political economist Patrick Bond has described the "granting, via a corrupting tender process, of small contracts to fledgling African entrepreneurs" in South Africa, a practice devised by the ruling African National Congress and termed "tenderpreneurship." These tenderpreneurs lobby for nuclear plants from Russia. For its part, the ANC loves these "big acquisitions from foreign governments, because it can charge rent from the investor, some of which invariably finds its way back into party coffers," journalist Tim Cohen points out.

Support for nuclear power extends to provincial and state levels of governments too. Several Canadian provinces have been at the forefront for small modular reactors (see chapter 6), with New Brunswick putting in more than C$30 million into two companies.[15] In China, governments of provinces like Hunan, Hubei, and Jiangxi have attempted to pressure the national government to build nuclear plants in their provinces. Part of their motivation has been the economic benefits that flow from these reactors that are paid for by national-level state-owned enterprises.

Singing nuclear power's "virtues"

Governments also act as advertising agencies for nuclear power, promoting it as desirable. Such advocacy not only serves as rationalizations for their own investment in nuclear power, but also effectively undermines opposition, either from local groups or from other elite constituencies.

Over the last decade, the most common advertisement has been nuclear energy's purported climate-mitigating properties. But these advertisements also come with the equivalent of "buy one, get one free" offers. An example is a public letter from the heads of state of the Czech Republic, Romania, France, Slovakia, Hungary, Slovenia, and Poland to the European Commission calling for nuclear power to be included as part of the EU climate and energy policy strategy. The letter was all about the merits of nuclear power for climate mitigation, asserting that it offers an "indispensable contribution to fighting climate change." To reinforce this point, the letter reiterated several unsubstantiated claims made by nuclear advocates. Since the EU has been relatively ahead in developing renewable technologies, the letter included claims about a "breadth of yet unexploited synergies between the nuclear and renewable technologies," talking about how "it guarantees the continued renewable deployment to much higher penetration levels." It then brought in the other silver bullet that has become a staple of nuclear advocacy in recent years: hydrogen. (Though it might be desirable, producing, transporting, and substituting current fuels with hydrogen is challenging as it is, even if nuclear reactors were not being used. Nuclear hydrogen amounts to two fantasies for the price of two.) To round these off, the letter also promised that nuclear power "generates a considerable number of stable, quality jobs, which will be important in the post-COVID recession."

The sad irony underlying the letter to the EU is that many of these heads of state are known for their commitment to fossil fuels and reluctance to adopt climate-mitigation policies. It reminds one of the Turkish proverb: "The forest was shrinking. But the trees kept voting for the Axe, for the Axe was clever and convinced the Trees that because his handle was made of wood, he was one of them."

It is not unusual for governments closely tied to fossil fuel interests to advocate for nuclear power. In Canada, where I live, some of the loudest expressions of support for nuclear power come from Alberta and Saskatchewan, the two largest producers of oil and the two provinces who generate their electricity overwhelmingly (over 80 percent in 2019) from fossil fuels. Talking about future nuclear power, especially using paper technologies like advanced nuclear reactors, is a strategy to shift attention away from their current energy mix.

In the case of Alberta, this is the third round of government talk about using nuclear technology to extract fossil fuels from its oil sands. The earlier phase was during the first decade of this century, amid fanfare for a so-called nuclear renaissance. But that idea was dead by 2011. Had it materialized, the earliest phase of intertwining nuclear technology and fossil fuels could have been the most spectacular—and dangerous. In *Stupid to the Last Drop*, investigative journalist William Marsden described how during the late 1950s the Alberta government supported Project Oil Sands, a plan to explode hydrogen bombs in order to boost oil production.

Climate mitigation is not the only rationale on offer. Below, in brief, I describe how government officials reproduce three of the talking points often utilized by nuclear advocates.

Necessity

Nuclear propaganda frequently resorts to the argument that there is no alternative to nuclear power to meet energy demands, to replace the public's fear of reactor meltdowns with a fear of blackouts. This is the "lights will go out" gambit I mentioned in chapter 3.

This argument is naturally resonant in developing countries, where blackouts are not uncommon to start with. Thus, for example, Bangladesh's science and technology minister justified a deal with Russia to import two reactors by arguing that the purpose was "to ease the power crisis that hampers our economic activities." In South Africa, a nuclear project management company explained: "The Cape provinces need large scale reliable power, and the only option is nuclear."

But this argument is mobilized to support building nuclear reactors in industrialized countries as well. For example, the UK energy secretary Ed Davey declared in 2013: "If people at home want to be able to keep watching the television, be able to turn the kettle on and benefit from electricity, we've got to make these investments. It's essential to keep the lights on and to power British business." The "keeping the UK's lights on" talking point matched what was put out by the Hinkley Point C media team. The same fear was mobilized in California, amid a heat wave in September 2022, by Governor Gavin Newsom to promise PG&E a $1.4 billion forgivable loan.

These arguments just do not square with the reality of a shrinking nuclear contribution to electrical energy. In the United Kingdom, nuclear power contributed 14.7 percent of the electricity produced in 2022, down from 26.9 percent in 1997.[16] In the meanwhile, renewables have risen over the same period, from 0.8 percent in 1997 to 41.5 percent in 2022. Likewise, in California, nuclear reactors produce roughly 8.5 percent of the total electricity generated within California in 2022, whereas renewable energy sources, without including large hydro dams, contributed around 35 percent.

A related tactic deployed regularly is using long-range projections of energy demand, based on unrealistic assumptions, to predict energy shortages. This unrealistic prediction is then coupled with a dismissal of alternative sources of energy to make a case for nuclear power being the only option. For example, in 2014, Jordan's energy minister claimed that electricity demand will triple by the year 2030 to make a case for acquiring reactors. In 1975, the International Atomic Energy Agency came up with two forecasts, both showing greatly increased demand on the island of Java for the period from 1986 to 1996, to argue for building nuclear power plants.[17]

But such projections are never realized. Indonesia's energy demand in 1996 was significantly lower than even the IAEA's low projection. Data from the International Energy Agency shows a modest 15 percent increase in Jordan's electricity consumption between 2014 and 2021, nowhere near the pace needed for a threefold increase. Other examples of this phenomenon are Turkey and Ghana.[18]

Jobs

Job creation has been a talking point for the industry for a while now. In 2010, for example, a Westinghouse advertisement announced that the "nuclear energy renaissance has already created thousands of new jobs. By providing reliable and affordable electricity, nuclear energy helps keep business competitive and power future worldwide job growth." More recently, the nuclear industry and its supporters have regularly argued for supporting the development of small modular reactors, because of their supposed potential for job creation.

The jobs argument is seen as having bipartisan appeal. Maria Korsnick, CEO of the Nuclear Energy Institute, the industry's lobbying organization, revealed to *Bloomberg* in 2017: "If you look at what we're passionate about, I would say jobs, jobs, jobs ... It's going to resonate very well with the current administration." The "current" administration then was that of Donald Trump, whose views about climate change were known well enough for the nuclear industry to pivot to a different argument. Around the same time, another US lobbying organization, Nuclear Innovation Alliance, claimed that "a U.S. SMR industry could create or sustain hundreds of thousands of American jobs." Implicit in this claim is the pitch to people like Donald Trump, so that he could take credit for these putative jobs at election time.

This pitch about nuclear projects creating lots of jobs has been borrowed uncritically by government officials. In 2013, for example, Prime Minister David Cameron talked up the deal with EDF as allowing for "the creation of 25,000 jobs, which is brilliant news for the South West and for the country as a whole." That number came directly from EDF.

Other countries have adopted the same jingle. For example, in 2017, an official from Turkey's Energy and Natural Resources Ministry announced, "About 10,000 people will be employed while the Akkuyu NPP's construction is most intensive, and about 3,500 jobs will be provided during operation. The majority will consist of Turkish citizens." Akkuyu is Turkey's first nuclear power plant, built by Russia. Russian officials added even more enticement by promising not only the involvement of "thousands of professionals" in Akkuyu but also the prospect of exports: "Turkish companies will gain relevant experience to participate in tenders for the construction of nuclear power plants in different countries."

The last promise is unlikely to hold. In the "different countries" that Rosatom is courting, nuclear power has to be sold by promising "thousands" of manufacturing jobs to their citizens.

The same jobs cannot materialize in both Turkey and these putative customer countries.

It is not just nuclear plants. Countries like Canada and Australia that have large reserves of uranium emphasize mining and processing jobs, often using exaggerated projections of nuclear energy expansion. In 2011, Australia's then prime minister, Julia Gillard, pushed a change in export policy to India, arguing that India would generate 40 percent of its electricity with nuclear energy by 2050 and "having access to this market is good for Australian jobs." A decade later, aside from a small shipment of uranium sent for testing in 2017, no uranium has left Australia for India. Nuclear power has never constituted more than 3.2 percent of India's electricity supply and is unlikely to amount to much more.

The emphasis on jobs should not be taken at face value. As Noam Chomsky explained with his usual sardonic humor in 1993:

> One can ... appreciate the passionate concern expressed by political figures, corporate leaders, and their press agents over the need to create jobs for suffering Americans. Heartening indeed, until we recall our lessons. Looking a bit more closely, we find that the word "jobs" has taken on an entirely new meaning: "profits." Thus when George Bush takes off to Japan with a bevy of auto executives in tow, he waves the banner "jobs, jobs, jobs," meaning "profits, profits, profits," as a look at his social and economic policies demonstrates without equivocation.[19]

Hypocrisy aside, the argument about nuclear power providing jobs is also misleading on at least three counts. First, any kind of investment, whether in nuclear power or shoe manufacturing, will lead to jobs, at least in the short term. The crucial question is how many jobs are created per unit of investment in each of these ventures. The academic literature is clear that nuclear power generates fewer jobs than renewables like solar and wind energy per unit of energy generated. Jobs in the renewable sector also tend to be more geographically distributed, whereas nuclear jobs are highly concentrated.[20]

Second, these jobs are sustainable only if the investment is toward a product that is commercially viable. Investing in an unviable or defunct technology that has no demand will not lead to long-term jobs. Building a factory to make audio cassette tapes might have made sense in the 1970s but not in the 2020s. Because electricity from small modular reactors will be far costlier than other forms of power (see chapter 6), a factory to manufacture these will not lead to jobs over and above the ones involved in building the facility.

Third, even if nuclear power might have created lots of jobs in the past, this is unlikely to be the case for future nuclear reactors. For decades, the nuclear industry has focused on reducing the numbers of workers needed, both to manufacture and operate power plants. This is unsurprising, because the main economic challenge faced by owners of nuclear plants is cost.[21] This explains proposals for nuclear reactor designs operating in a completely automated fashion, or with minimal operators. Likewise, the nuclear industry has switched to using modular methods to manufacture reactor components, to reduce the amount of labor needed to build a nuclear plant. This trend is only intensifying, further directing investment away from labor toward capital.

Modernity

In developing countries, nuclear power is also often advertised using an older selling point: a marker of modernity or development. Such ideas of modernity and advancement appeal especially to elite constituencies, as do claims about their country having achieved some kind of great status. Nuclear power allows government leaders to score on these points.

Talking about the uniqueness of nuclear power also evidently boosts the status of the leader or the government, as the entity that put in place these achievements.

Consider the case of Turkey, one of the countries that is constructing its first nuclear power plant (as of the time of this writing). This construction has played very well into President Recep Tayyip Erdoğan's brand of politics, which relies on portraying Turkey as a great state with him at its helm.[22] One manifestation has been yoking nuclear power to the historical occasion of the founding of the Turkish republic. In Erdoğan's words, "In 2023, we will commission the first reactor at this plant, and Turkey will thus join the countries that use atomic energy. In 2023, we will mark the 100th anniversary of our republic with the successful completion of this project." Despite this keen interest, Akkuyu has not been commissioned as of January 2024.

Note also how Erdoğan marks off nuclear power as the exclusive preserve of some nations. Similar messages are offered by other government functionaries and politicians. At a 2015 launch function for the Akkuyu plant, Turkey's energy minister Taner Yıldız proclaimed, "Development cannot take place in a country without nuclear energy." And in 2007, an AKP parliament member, Mustafa Ozturk, stated: "Nuclear power plants reflect the strength, the level of development, and the prestige of a country. We have been late for 40 years in shifting to nuclear technology, thus, we have to be successful in bringing this high-tech to our country."

For another example, we can go to China. In a November 2017 meeting of the Chinese Society for Electrical Engineering, Shi Lishan, deputy director of the Nuclear Power Division of the National Energy Administration, announced, "Nuclear power is a symbolic industry and indicator of a country's industrialization and modernization, as well as a comprehensive manifestation of the country's level of processing and manufacturing, social management capabilities, and safety management. Large countries must have a nuclear industry" (according to Google Translate).

Such sentiments are not exclusive to developing countries. In Canada, the government department responsible for promoting nuclear technology proclaims on its website: "Nuclear energy technology is a hallmark of the world's leading industrial nations." While French President Emmanuel Macron averred, "Everything that makes France an independent, listened-to and respected power is based on the nuclear industry."

Whether nuclear power should have been considered a sign of modernity in the 1950s, and a desirable one at that, can be debated. But today, seventy years after the first nuclear power plants started functioning, this argument is stale at best. Yet it continues to be deployed by numerous leaders around the world.

Nurturing, supporting, and cheering

Explaining why countries like China or Russia might be constructing nuclear power plants is not so difficult, for these countries don't value market competition, and the government can push whatever technology it desires. But there seems something very odd about the role of the state in countries such as the United States or the United Kingdom, which supposedly favor market competition and not picking winners and losers. But as I have outlined here, the governments of these countries have played a big role in developing and promoting nuclear energy. And this support is not limited to just the early stages, when technology developers must be allowed to mature before being put on the proverbial "level playing field." But in the case of nuclear power, this early stage has long gone, and we are dealing with a technology that has more than matured—perhaps to the point of obsolescence.

Heavy state intervention has become more common over

the last few decades, but only in ways that allow the owners of capital to make more profits. These profits, of course, don't materialize from thin air, but by short-changing members of the public. In the case of the Hinkley Point deal, the UK House of Commons Committee of Public Accounts was quite blunt: "Consumers are locked into an expensive deal lasting 35 years … [They] are left footing the bill and the poorest consumers will be hit hardest. Yet in all the negotiations no part of Government was really championing the consumer interest."

Nuclear proponents are not shy about demanding such benefits for investors. An August 2022 report from the Dalton Nuclear Institute at the University of Manchester says explicitly that "the state" should be "creating an environment in which the private sector is willing to make the huge capital investments associated with delivering nuclear energy." What might such an environment look like? While the details vary from place to place and with time, the broad contours involve the government easing regulatory requirements, especially those associated with safety of the nuclear plant and its environmental impact, and customers or taxpayers paying for constructing and operating highly priced nuclear plants, and footing the bill for decommissioning reactors and managing radioactive waste.

Many of these have already been put in place for Sizewell C, the reactor project to which Boris Johnson funneled £700 million. The RAB model of charging customers ahead of electricity production puts ratepayers on the hook from the get-go. The UK safety regulator had essentially signed off on the project in July 2022. The precedent for British taxpayers paying for decommissioning reactors and managing the radioactive waste has already been set. All that is left is for investors to come in with their capital. And the British government is doing its bit to woo them, starting with corporations and funders in the United Arab Emirates, Australia, and Saudi Arabia.

Government officials and political leaders also mobilize several arguments in favor of nuclear power, arguments that are worn smooth with repetition. Depending on the context, they can choose from a standard menu that nuclear advocates the world over draw upon: delivering energy security, saving us from climate change, providing jobs, and making a country modern. In the case of Sizewell C, the leading items on the menu are the first two. As a Department for Business, Energy, and Industrial Strategy official put it, "The Government is fully committed to boosting our domestic energy security through nuclear power."

The sincerity underlying these statements is dubious at best. A great example is President Donald Trump. Despite his well-known climate denialism, he was greatly supportive of nuclear energy. In June 2017, at an event called "Unleashing American Energy," he proudly announced "six brand-new initiatives to propel this new era of American energy dominance," then elaborating, "First, we will begin to revive and expand our nuclear energy sector—which I'm so happy about—which produces clean, renewable, and emissions-free energy." As one might expect from someone who cared little for facts, he did say "renewable."

Except for the nebulous status of being modern, the other promises can be measured. And history tells us that these promises will not be realized. But this history is brushed aside with the claim that this time it will be different. And by the time such problems become apparent, the ruling leaders at the time of the investment decision have moved on, either into retirement or into other lucrative private-sector roles.

In the next chapter, I highlight one important reason for governments to support nuclear energy: its close connections to the nuclear weapons enterprise.

5

May the Atom Be a Soldier: Nuclear Power for War

Nuclear power must not only be safe but must also be used solely for peaceful purposes.

International Atomic Energy Agency

And Denial Is Not a River in Egypt.

Anonymous graffito, 1986

When I was a researcher at Princeton University, I inherited an old couch. Although a bit shabby and rickety, the couch was cozy and allowed me many hours of reading books stretched out in comfort, away from my desk. But more than that luxury, I cherished the couch because it belonged to the legendary Ted Taylor.

Ted is perhaps the smartest US nuclear weapons designer ever. Between 1949 and 1956, he worked at the birthplace of nuclear weapons, the Los Alamos National Laboratory in New Mexico. Ted designed the generation of weapons that came after those dropped over Hiroshima and Nagasaki. He was responsible for the lightest and smallest fission bomb ever made, at just fifty pounds (the Davy Crockett), and the largest-yield fission bomb ever exploded (the Super Oralloy Bomb). In between, he registered for a PhD program at Cornell University in 1953 and received a doctorate in theoretical physics in 1954.

Restless and brilliant, Ted obsessively worked on two things: producing nuclear explosions and controlling them. In 1952, he even created "the world's first atomic cigarette lighter," points out Richard Miller in his book *Under the Cloud*. Ted had focused the flash of light produced during a nuclear weapons explosion to light a Pall Mall cigarette. Ted later went on to join Project Orion, the futile effort to use nuclear weapons explosions to propel a spacecraft for interplanetary or even interstellar journeys.

Things changed in the mid-1960s after a stint at the US Department of Defence, where Ted began to see how nuclear technology could be abused by governments, terrorists, and criminals—which aren't separate categories in my mind. He began to walk on a decidedly different path, transforming himself from a nuclear weapons designer to a passionate opponent of nuclear weapons and nuclear energy.[1]

By the 1970s, Ted's moral convictions against all nuclear technology were clear: "If it were possible to wave a wand and make fission impossible—fission of any kind—I would quickly wave the wand," he told journalist John McPhee. McPhee accompanied Ted to various nuclear installations in the United States and eventually wrote *The Curve of Binding Energy*, which was nominated for a National Book Award. "I have a total conviction—now—that nuclear weapons should not be used under any circumstances. At any time. Anywhere. Period. If I were king. If the Russians bombed New York. I would not bomb Moscow," Ted told McPhee.

Thanks to his work on, and clear commitment to, nuclear disarmament, Ted was a legend in my circles. This is why I felt privileged to possess his couch.

Once Ted understood the existential dangers associated with nuclear weapons, supporting nuclear disarmament was obvious. But what explained his switch to opposing nuclear energy?

I ask this question partly because of my own background. Though I can claim none of Ted's brilliance, and I definitely took much longer than he did to complete the degree, I also received a PhD in theoretical physics. While I was growing up in India, the virtues of nuclear energy had been constantly dinned into my ears. India's Department of Atomic Energy extensively promoted its visions of cheap electricity from nuclear reactors, and there was never any doubt in the minds of my schoolmates about the importance of atomic energy for economic development and eliminating poverty. I imagine Ted was exposed to a similar barrage of propaganda in the 1950s. At the same time, I knew enough about the horrors that befell Hiroshima and Nagasaki—although I can't quite recall where I learned about that—to have a moral and ethical aversion to nuclear weapons for as long as I can remember.

If you had asked me about my views when I was an undergraduate physics student at the Indian Institute of Technology in Kanpur, odds are that I would have said, "Nuclear weapons bad; nuclear energy good." It was only years later that I started becoming critical of nuclear energy, realizing that the costs and risks associated with the technology outweigh any benefits. The accident in Chernobyl definitely played a role, as was a gradual realization of the environmental and health impacts of a range of modern technologies.

This background about me might help explain why I was curious about what led Ted to oppose nuclear energy. His answer had little to do with the economic costs or environmental impacts but related to the capacity of any nuclear reactor to provide materials to make weapons. As he explained in a talk to the Nuclear Age Peace Foundation in 1996, the constructive and destructive uses of nuclear technology were so intimately related that "the benefits of the one are not accessible without greatly increasing the hazards of the other."

These close connections are precisely why nuclear power is attractive to some proponents. The overlap between the two pursuits explains why some governments support nuclear power, despite the manifest shortcomings of the technology detailed elsewhere in this book.

The connections work both ways. Nuclear power can serve as a gateway to a variety of military applications. Conversely, military nuclear technologies can propel the development of a nuclear energy industry. Both pathways help cement the political power of institutions involved in the development of these technologies and make it harder to exert any kind of democratic control over them.

This chapter describes the long history of these connections. Such connections are on display in countries that first developed nuclear weapons (e.g., the United States), and in countries that first set up nuclear power plants (e.g., India). I also describe how private corporations profit from both technologies. There is a parallel history of the nuclear energy industry denying the connections between nuclear power and weapons—except when nuclear energy operators are experiencing financial trouble. Finally, I document the recent spurt in a variety of prominent actors emphasizing the connection between the two technologies to advocate for government support to the nuclear industry.

A historical relationship

In 1946, the US government set up a committee headed by two influential individuals: Dean Acheson, who was to subsequently become the US secretary of state, and David Lilienthal, then chair of the Tennessee Valley Authority. The resulting *Report on the International Control of Atomic Energy* warned that "the development of atomic energy for peaceful purposes and the development of atomic energy for bombs are in much of their course interchangeable and interdependent." That fundamental fact has not changed since then.

As the Acheson-Lilienthal report makes clear, the oldest military application for nuclear technology is the nuclear weapon. The first reactors constructed in many countries were all intended to produce plutonium to be used in nuclear weapons.[2] The Hanford site in the US state of Washington was home to the first nuclear reactors that produced plutonium for the bomb dropped on Nagasaki on August 9, 1945.

The next major military applications involved various nuclear-powered marine vessels. The most prominent of these are nuclear-powered submarines. But there are also nuclear-powered aircraft carriers, nuclear-powered cruisers, and nuclear-powered icebreakers. Besides changing the nature of warfare, nuclear-powered propulsion ended up shaping the nuclear energy industry too. The reactor designs first developed to power nuclear submarines became the basis for nuclear power plants in many countries around the world.

In the United States, the driving force behind the choice of the reactor design for submarines was Admiral Hyman Rickover. Rickover is often described as the "father of the nuclear navy" because he pioneered the use of nuclear propulsion in submarines. After exploring different kinds of nuclear reactors (for example, sodium-cooled fast neutron reactors), he selected a design known as the pressurized light water reactor to power the first nuclear submarines. This reactor type uses purified water to carry away the heat produced when the uranium fuel undergoes fission reactions. Rickover's selection of this design for the US naval submarine program meant that as the United States built up its nuclear submarine force, many reactors of this design were manufactured. As a 2020 analysis from the Idaho National Lab shows, the repeated manufacture meant that many of the problems with the initial versions of the design were overcome, at government expense. Other reactor designs that had their share of problems did not have this advantage.

Over the next three decades, this design ended up dominating the nuclear power plant market, beating out various other alternatives that were proposed in other countries, and even designs proposed by rival developers in the United States. Had this design not been used to produce nuclear-powered vehicles armed with nuclear weapons patrolling underwater around the world, pressurized light water reactors might not have had the large market share they do today (304 of the 413 reactors listed as operational as of January 2024 by the International Atomic Energy Agency's Power Reactor Information System database). The relationship between nuclear weapons and nuclear energy was no secret, especially at the inauguration of the atomic age. Concerns about the potential spread of nuclear weapons were high during that period. Such concern reached paranoid levels in the United States, the sole possessor of nuclear weapons, perhaps because of its intimate knowledge of the destructive capabilities of these weapons.

Perspectives from customers

The interdependence of these two technologies was laid out in a March 1987 interview to the *Washington Post* by Pakistan's dictator, General Muhammad Zia-ul-Haq: "Once you have acquired the technology, which Pakistan has, you can do whatever you like. You can use it for peaceful purposes only; you can also utilize [it] for military purposes." General Zia went on to reassure his interviewer, claiming, "We have never said we are incapable of doing this. We have said we have neither the intention nor the desire." But we know, with the wisdom of hindsight, that the latter assertion was not true.

We may be seeing a repeat of this two-step move in Saudi Arabia. Around a decade ago, Saudi officials announced that the country is embarking on an ambitious energy diversification plan, including a massive addition of nuclear power. Soon Saudi officials had announced plans to install 18,000 megawatts of nuclear generation capacity, equivalent to over fifteen large nuclear power plants, by 2032. The announcement was welcomed by the nuclear industry, which was reeling from the impact of the Fukushima accident.

A royal decree from April 2010 reasoned that "the development of atomic energy is essential to meet the Kingdom's growing requirements for energy to generate electricity, produce desalinated water and reduce reliance on depleting hydrocarbon resources." There is something odd about the claim that atomic energy was *essential* for Saudi Arabia. The natural and obvious way to reduce reliance on hydrocarbons in sun-baked Saudi Arabia is solar energy. By 2010, the cost of solar power plants had already started declining. So, why atomic energy?

The answer had been floating in the air for years, but the definitive confirmation came in March 2018. That, too, from the most powerful individual in the country, Prince Mohammed bin Salman (or MBS, initials that some would say stand for Mister Bone Saw, after what was done to journalist Jamal Khashoggi, once he fell out of favor with the prince). MBS painted the context for the Saudi interest in nuclear technology in an interview with *CBS News*:

MBS: Iran is not a rival to Saudi Arabia. Its army is not among the top five armies in the Muslim world. The Saudi economy is larger than the Iranian economy. Iran is far from being equal to Saudi Arabia.

CBS: But I've seen that you called the Ayatollah Khamenei, "the new Hitler" of the Middle East.

MBS: Absolutely.

CBS: Why?

MBS: Because he wants to expand. He wants to create his own project in the Middle East, very much like Hitler who wanted to expand at the time. Many countries around the world and in Europe did not realize how dangerous Hitler was until what happened, happened. I don't want to see the same events happening in the Middle East.

CBS: Does Saudi Arabia need nuclear weapons to counter Iran?

MBS: Saudi Arabia does not want to acquire any nuclear bomb, but without a doubt if Iran developed a nuclear bomb, we will follow suit as soon as possible.[3]

Might the desire to be able to develop a nuclear bomb "as soon as possible" be why Saudi Arabia is interested in nuclear energy? Will Saudi Arabia use its proposed acquisition of nuclear power plants in some way to make nuclear weapons? For that matter, might Egypt or Belarus or Turkey, which have been building their first nuclear power plants, eventually acquire nuclear weapons? We can't say for sure, but as history shows, nuclear reactors can only help.

The illusion of separation

Although many of these connections between nuclear energy and nuclear weapons were clear from the beginning of the atomic age, the nuclear industry and advocates for nuclear power have tried systematically to erase this connection. They portray nuclear energy as peaceful and beneficial to humanity and nuclear weapons as an undesirable goal, or a necessary evil at best. A clear if pithy illustration of this desire to turn away from nuclear weapons toward nuclear energy is the Soviet slogan "May the atom be a worker, not a soldier."

Historically, the most significant effort to separate nuclear energy from weapons was President Dwight Eisenhower's Atoms for Peace program. Speaking at the United Nations General Assembly in December 1953, President Eisenhower argued that "if the fearful trend of atomic military buildup can be reversed, this greatest of destructive forces can be developed into a great boon, for the benefit of all mankind" and went on to lay out various steps in order to "hasten the day when fear of the atom will begin to disappear from the minds of the people and the governments of the East and West."

A half century later, the desire expressed by Eisenhower had turned into downright denial. During the height of the hype about the nuclear renaissance in the first decade of this century,

Anne Lauvergeon, then the head of France's Areva, asserted in a 2009 article in *Daedalus* that the growth of nuclear electricity does "not equate—and should not be equated—with increasing proliferation risks."

Advocates also focus on the differences between the two technologies while simultaneously caricaturing the argument made by those who see a relationship between the two technologies. An op-ed by Ted Nordhaus, co-author of *An Ecomodernist Manifesto*, illustrates this strain of argument by asserting that "the tendency to conflate nuclear energy with nuclear weapons" is "extremely misleading." Nordhaus contends "Neither the physics nor the technologies are the same, nor are the institutions that manage the two technologies."

Nordhaus's characterization is disingenuous at best. No one claims that a nuclear power plant works the same way as a nuclear bomb. As described later, however, there are clear material and institutional connections between nuclear energy and weapons. Indeed, in the United States, where Nordhaus is based, the same institution, the Department of Energy, manages the production of nuclear weapons and funds research and development of nuclear energy technologies.

Whether it is Eisenhower or Nordhaus, promoters of nuclear power feel that it is important to create the perception of a clear divide between the two technologies. Otherwise, they feel, those who oppose weapons of mass destruction will also oppose nuclear energy.

This diagnosis was explicitly laid out in a 1984 report from the US Office of Technology Assessment. The political context for this report was marked by two oppositions: opposition to the Reagan-era nuclear weapons buildup by the United States and the Soviet Union, and opposition to nuclear energy in the aftermath of the 1979 Three Mile Island accident. The report's conclusion: "Public acceptance of nuclear power cannot be expected to increase substantially until the two nuclear technologies *are separated in people's minds*" (my emphasis). In other words, the public should ideally not understand the relationship between nuclear weapons and energy.

A volte face

After decades of maintaining that there is no connection between nuclear energy and nuclear weapons, some nuclear power advocates have suddenly switched their argument in recent years. They now explicitly invoke the connection between the two technologies to counteract the economic pressures on nuclear power described elsewhere in the book.

Let me offer one illustration. In December 2020, the Office of the President of France issued a statement providing three motivations for continuing to support its nuclear energy sector: "Our energy and ecological future depends on nuclear power; our economic and industrial future depends on nuclear power; and France's strategic future depends on nuclear power."

Why would Macron make this public declaration of faith in the nuclear industry? At that time, France's nuclear industry had been failing, and failing badly at that. Its top nuclear design company (Areva) and its major nuclear construction company and utility (Électricité de France, some of whose activities are described in chapter 4) had been struggling for nearly fifteen years with constructing power plants featuring their flagship European pressurized reactor (EPR) design. Construction in Flamanville, France, and Olkiluoto, Finland, had been abject disasters, running over a decade late and billions of euros over budget. The experience of building in Taishan, China, the world's fastest-growing nuclear market, was no success either, and it was clear that China would never again purchase a French nuclear reactor. Apart from the Hinkley Point C project in the United Kingdom, France had no orders for its nuclear reactors.

France prided itself on the heavy reliance on nuclear energy. But the *World Nuclear Industry Status Report 2020* documents the share of the country's electricity from nuclear reactors declining, from 78.5 percent in 2005 to just over 70 percent in 2019; not one French reactor was ranked among the top 100 units in the world. Électricité de France faced major debt problems and, in July 2020, the national Cour des Comptes (Court of Accounts) had publicly criticized the agency's project management skills.

This was the context for Macron's defense of the nuclear industry. And the argument he used was to emphasize the military and strategic value of nuclear power.

France was not unique. Later in this chapter, I detail how nuclear proponents in the United States and the United Kingdom highlight the overlap between nuclear energy for electricity generation and for military purposes.

Connections

What are the linkages between nuclear energy and the capacity to make nuclear weapons? One might discern three separate, even if interrelated, elements having to do with technical overlaps, interchangeability of personnel, and institutional imperatives.

I will start by describing how the infrastructure to produce nuclear electricity also offers the technical capacity to advance the ability of countries to make nuclear weapons materials. My colleagues at Princeton were among the pioneers in exploring this connection, and this is in part why Ted Taylor ended up there. One of my colleagues at the school, Hal Feiveson, coined the term "latent proliferation" in his 1972 PhD thesis. What Hal described was how a country with nuclear power plants arrives at a point that is "short of the actual possession of nuclear weapons, but that can account for much of what has to be done technically to acquire them," as summarized by Ted in his 1996 lecture.

Technology

Uranium occurs naturally in two main varieties, called isotopes. Isotopes are two or more forms of the same chemical element that have different masses. The difference in masses traces back to the different numbers of neutrons in different isotopes of an element. All isotopes of a given element have the same number of protons in their nuclei. Because atoms are electrically neutral, the number of protons equals the number of electrons, which determines all the chemical properties. Consequently, all isotopes will have identical chemical properties.

The element uranium, for example, has ninety-two protons. The heavier isotope, uranium-238, has 146 neutrons, but the lighter one, uranium-235, has only 143 neutrons. In uranium mined from the ground, for every thousand atoms of uranium, the overwhelming majority, 993 atoms, will be uranium-238. The remaining seven will be uranium-235.

The uranium-235 isotope has a special property: it can sustain a chain reaction, which is the basis of both nuclear power plants and nuclear bombs. When a uranium-235 nucleus is struck by a neutron, it splits (fissions), producing other lighter radioactive elements and neutrons. This fission occurs regardless of the neutron's energy level. If these neutrons can go on to trigger further fission reactions, the result is a chain reaction. Materials capable of sustaining a chain reaction are called fissile materials. Besides uranium-235, the most commonly used fissile material is plutonium-239.

The concentration of uranium-235 in nature is usually too low for such a chain reaction to occur. Therefore, to make nuclear weapons the uranium-235 concentration must be enriched, from 0.7 percent to ideally around 90 percent. This is done in uranium-enrichment plants. The dominant technology today to enrich uranium is the centrifuge, a machine that spins around at high speeds, similar to a washing machine, causing the two different isotopes to move away from each other.

Nuclear reactors are more varied, and some can operate with natural uranium without enrichment. But many nuclear reactor designs, in particular the pressurized water design first developed for nuclear submarines and then widely adopted in nuclear power plants, require uranium-235 concentrations in the range of 3 to 5 percent.

The connection between nuclear energy and nuclear weapons derives from the fact that any technology capable of enriching uranium-235 from 0.7 percent to 3 percent can further enrich it, even up to the levels of concentration needed to build nuclear weapons. This potential ability to use uranium-enrichment technology for making either nuclear fuel or nuclear weapons materials was the underlying technical reason for concern about Iran's centrifuge program.

Enriching uranium is one route to making a nuclear weapon, such as the bomb that destroyed Hiroshima. A second route to the bomb was demonstrated to the world in 1945 at Nagasaki. The Nagasaki bomb used the other fissile material, plutonium. Plutonium is not found in nature but is produced from uranium. When uranium fuel is irradiated in the reactor, the uranium-238 isotope absorbs neutrons and gets transmuted into plutonium-239. This plutonium can be separated from uranium and other chemicals in the irradiated fuel through a chemical operation called reprocessing.

Just as centrifuges can be utilized to enrich uranium-235 to concentration levels that are either only sufficient for use in nuclear power plants or high enough for making nuclear weapons, separated plutonium can be used for making both bombs and fuel for nuclear power plants. What a country does with the plutonium is its decision, and countries can begin producing plutonium ostensibly for electricity generation and then use it to make or test nuclear weapons. Examples of countries that have done this are India and North Korea.

In contrast to enrichment, wherein uranium needs *more* enrichment to be used in nuclear weapons, plutonium typically used in nuclear weapons derives from uranium that has been *less* irradiated—in other words, left inside the reactor for a shorter period of time. It is not just uranium-238 that can absorb a neutron inside reactors, but also plutonium-239, which then gets converted into the isotope plutonium-240. The latter is not so desirable for use in nuclear weapons. Weapons designers prefer using plutonium with less than around 6 percent of plutonium-240.

This arcane technical detail is relevant because nuclear energy advocates often argue that plutonium derived from commercial nuclear power plants cannot be used in nuclear weapons.[4] Nuclear power plants are optimized to extract as much energy as possible from a given amount of uranium; as a result, the plutonium-240 concentrations tend to be much higher, around 25 percent in the case of light water reactors.

Preference is not the same as necessity. It is not as though when the fraction of plutonium-240 exceeds 6 percent, the material could not be used to build nuclear weapons. As the US Department of Energy announced: "Virtually any combination of plutonium isotopes ... can be used to make a nuclear weapon." The DOE should know, having overseen over a thousand nuclear explosions. In other words, nuclear power plants can also be sources of plutonium for nuclear weapons. More generally, even though there are technical differences between different kinds of nuclear reactors, they all can be used to make nuclear weapons materials.

Finally, there is also the connection with nuclear-powered aircraft carriers and, especially, nuclear-powered submarines. These vehicles are typically used to carry nuclear weapons and launch them when called upon to do so. The material used to fuel submarine reactors is typically highly enriched uranium.

Some countries choose to build nuclear submarines even though they don't have nuclear weapons. Brazil, which has signed the Nuclear Non-Proliferation Treaty (NPT) as a non-nuclear state, has a nuclear submarine program. Even this military connection has served to further the ability of its nuclear power lobby to obtain state support.

If I had a dollar for every time I have been told about so-called thorium reactors in response to any of my criticisms of nuclear energy, I would be pretty rich. Certainly so if I could get a dollar for every comment underneath one of my online articles dismissing my arguments because I did not take into account thorium reactors. One of the arguments made in favor of these mythical beings is that they cannot be used to make nuclear weapons. I use the term "mythical" because there is no commercial thorium reactor in existence, and there are several serious technical problems with the thorium cycle that are yet to be solved. But even if one were to be built, it is a fallacy to think they cannot be used to produce materials for nuclear weapons.

Thorium itself, unlike uranium, cannot be used as reactor fuel because it is not fissile; it cannot sustain a chain reaction. Therefore, most so-called thorium reactors actually involve a fissile isotope of uranium, uranium-233, produced when the thorium-232 isotope absorbs a neutron and undergoes a series of nuclear decays. Uranium-233 is not found in nature, because it is unstable.

The isotope uranium-233 *can* be used to make nuclear weapons. In some respects, it is superior to both isotopes commonly used in nuclear weapons, uranium-235 and plutonium-239. Compared with weapons-grade uranium, less uranium-233 is needed to start a chain reaction—a lower critical mass in technical parlance. In other words, making a nuclear weapon requires a smaller amount of uranium-233 than uranium-235. Relative to plutonium-239, uranium-233 fissions spontaneously at a lower rate. A weapon made of uranium-233 is less likely than one made of plutonium to go off in advance of when the designers want it to. Thus, uranium-233 has some very desirable properties for those who wish to make nuclear weapons.

At the same time, uranium-233 does have a property that makes it less desirable. When uranium-233 is produced in reactors, it *usually* comes out in conjunction with another isotope of uranium, uranium-232, which is radioactive and emits high energy gamma rays. This is the main reason for nuclear weapons designers not preferring uranium-233. While undesirable, this property does not completely preclude the use of uranium-233 to make nuclear weapons, because it is possible to isolate pure streams of uranium-233. What is more, the admixture with uranium-232 makes it even less desirable to use as fuel in nuclear reactors, because workers would be subject high radiation doses. Thus, so-called thorium reactors would have more of a problem with the nuclear properties of uranium-233 than would bomb makers.

People

The second way that acquisition of nuclear power plants could contribute to a country's nuclear weapons quest is by providing a reason to train people in the many skills that are involved in both pursuits. Munir Ahmed Khan, chair of Pakistan's Atomic Energy Commission, who was responsible for launching Pakistan's nuclear weapons program, explained it quite plainly to scholar George Perkovich, now at the think tank Carnegie Endowment for International Peace:

> The Pakistani higher education system is so poor, I have no place from which to draw talented scientists and engineers to work in our nuclear establishment. We don't have [a] training system for the kind of cadre we need. But, if we can get France or somebody else to come and create a broad nuclear infrastructure, and build these plants and these laboratories, I will train hundreds of my people in ways that otherwise they would never be able to be trained. And with that training, and with the blueprints and the other things that we'd get along the way, then we could set up

separate plants that would not be under safeguards, that would not be built with direct foreign assistance, but I would now have the people who could do that. If I don't get the cooperation, I can't train the people to run a weapons program.[5]

Khan's predicament can be traced back to the state of Pakistan's scientific and technical resources when it became independent from two centuries of British colonialism in 1947. During the 1950s, as the country undertook its first economic-planning exercises, its planning board took the help of a set of experts from Harvard University. One of these experts, Maurice Kilbridge, undertook the first study on the economic viability of nuclear power in Pakistan, and concluded that the goal was unrealistic for the foreseeable future.

Kilbridge's 1958 report, "The Prospect for Nuclear Power in Pakistan," identified the major concern: "Probably not more than 10 persons in all Pakistan … have any extensive training in nuclear technology, and … not many more [have] the basic education necessary to absorb such training." Pakistan's way out of this predicament ended up being the US Atoms for Peace program. As part of this initiative, in 1955, the two countries signed a five-year Agreement for Cooperation on the Civil Uses of Atomic Energy, which included a training program for scientists and engineers.

The program proved critical for Pakistan's nuclear energy and weapons programs. By 1961, my former Princeton University colleague Zia Mian wrote, "the newly created Pakistan Atomic Energy Commission (PAEC) had 144 scientists and engineers, who either had already received training abroad or were currently being trained abroad. Among those trained in the U.S. was Munir Ahmed Khan."[6]

Pakistan was not the only country that had its nuclear personnel trained in the United States. Pakistan's neighbor India also benefited from US training with over 1,100 scientists and engineers going to the Argonne Laboratory and other facilities between 1955 and 1974. Nor was the United States the only country providing such training. Its neighbor Canada trained 263 Indian scientists and engineers prior to 1971, according to Bob Anderson, a pioneer in the study of India's nuclear history.[7] These exchanges between the United States and Canada and the two nuclear powers of South Asia were to stop after 1974. That was when Indian scientists and engineers, trained in many countries around the world, conducted India's first nuclear weapons explosion utilizing plutonium produced in the CIRUS reactor constructed by Canada, which also used heavy water from the United States. Well before this test, officials in the United States and, especially, Canada were concerned about the possibility that the people being trained could help develop nuclear weapons and that the nuclear facilities they were helping India and Pakistan set up could be used to produce plutonium usable in nuclear weapons. But they were overruled by other officials interested in supporting the nuclear industry, either for profit or for geostrategic reasons, who downplayed the risk of weapons acquisition. Similar considerations lead some to argue in favor of exporting nuclear power plants to Saudi Arabia today.[8]

But even after India's 1974 nuclear test, potential markets for nuclear reactors continued to be alluring. Nowhere was this more evident than at the Massachusetts Institute of Technology, where I was a postdoctoral fellow before I moved to Princeton. As detailed in a March 2007 article in the *Boston Globe*, thirty-five Iranian students arrived there in the summer of 1975 to study nuclear engineering. For the extra professors and classroom space, the Iranian government paid MIT more than half a million dollars.

The influx of students was occasioned by the Shah of Iran proposing to build over twenty nuclear reactors. In 1975, too, US Secretary of State Henry Kissinger signed a National Security Decision Memorandum, which laid the basis for the planned sale of nuclear reactors to Iran at an estimated cost of over $6 billion. Kissinger absolved himself to the *Washington Post* in 2005 by saying, "They were an allied country, and this was a commercial transaction. We didn't address the question of them one day moving toward nuclear weapons." Following the 1979 revolution, Iran's relationship with the United States changed, and so did the attitude of universities like MIT toward Iranian students interested in nuclear engineering.

This history was to be recounted in the aughts, following accusations, albeit unproven, that Iran was developing nuclear weapons. In its 2007 article, the *Boston Globe* traced many of these MIT students and found a number who worked in the Iranian nuclear program. One former student went on to become the deputy director of Iran's Atomic Energy Organization and another to run a nuclear research reactor in Tehran. The best-known MIT graduate was Ali Akbar Salehi, who became Iran's representative to the International Atomic Energy Agency.

Iran is a complicated case. Much of the discussion is based on statements from US or Israeli intelligence sources, which are hard to evaluate for someone on the outside. Nevertheless, what has been happening in Iran does show quite clearly that people who are trained ostensibly to build a nuclear energy program can then go on to design or build facilities—uranium centrifuges, heavy water reactors—that could be used to make nuclear weapons materials.

In countries possessing nuclear weapons, nuclear energy programs can serve a different purpose: providing jobs to people trained in the nuclear sciences and related fields as part of the weapons program. In the United States, there is a long tradition of employees of the "nuclear navy," which operates nuclear-powered submarines that constitute one of the three ways of attacking any part of the world with nuclear weapons, moving on to the "civilian" nuclear industry after retirement. Nuclear energy proponents argue that "prospective employment in the civilian nuclear power sector is a core incentive to academic training and military careers in nuclear energy" and "this supply chain of expertise is at least as essential as the material inputs."

Thus, the existence of over ninety nuclear power reactors in the country allows the US military to attract people to its nuclear-weapons-related segments. In addition, the Institute of Nuclear Power Operators, which offers advice to nuclear power plant owners on the safety of their facilities, is largely populated by former naval workers.

The nuclear navy provides a great springboard. Consider the case of William Ostendorff, a former navy officer who ended up serving as a commissioner at the US Nuclear Regulatory Commission for six years. Ostendorff worked in the navy from 1976 to 2002; during this stint, he worked on six submarines and commanded a nuclear attack submarine as well as a nuclear attack submarine squadron. Ostendorff also went into the administrative wing the National Nuclear Security Administration, which oversees the United States' nuclear weapons. In a 2019 podcast interview, he attributed his NRC appointment to precisely those factors.

Another example is Admiral John Richardson, a former chief of US naval operations appointed to Exelon Corporation's board of directors in September 2019. The press release announcing his appointment described his experience "leading the U.S. Navy and serving on the Joint Chiefs of Staff during two presidential administrations" as a "critical asset to our board ... His depth of experience in nuclear oversight and his operational expertise are invaluable." Why exactly this experience and expertise are valuable to the company might be inferred from the accompanying statement about him joining "Exelon's Finance and Risk and Generation Oversight Committees." For context, Exelon had been lobbying the state of Illinois to subsidize its nuclear reactors (see chapters 2 and 4).

In India, many officials involved in the development of nuclear energy have moved on to the military side as they became more senior. Placid Rodriguez, a metallurgist, joined the Department of Atomic Energy (DAE) as a twenty-year-old, working for decades within various institutions associated with the DAE. He ended up as the director of the Indira Gandhi Centre for Atomic Research and in charge of developing fast breeder reactors. When he retired, he was promptly appointed to the Defence Research and Development Organisation, which designs and manufactures the missiles that would deliver India's nuclear weapons.

Raja Ramanna, a leader of the team that oversaw the first nuclear weapons test in 1974, was appointed to a senior position for a few years at the Department of Defence in the late 1970s. In his autobiography, *Years of Pilgrimage*, he recalled: "The defence forces took well to my induction. They respected me, as I was from the Bhabha Atomic Research Centre (BARC), an organization about which they were well-informed and proud of ... Most importantly, the fact that I'd been involved in the development of a prototype weapon lent me a special status." In 1981, he moved back to the Department of Atomic Energy, becoming the director of BARC.

Such revolving doors between the military and civilian sectors are quite common in many nuclear countries.

Institutions

There is a third connection between nuclear weapons and nuclear energy: it is often the same institutions that lead work in both areas. The case of the US Department of Energy is a good example. Kenneth Bergeron's *Tritium on Ice* from 2002 points out that "its charter" involves both the "promotion of commercial nuclear power and production of nuclear weapons." The connection is common enough that in a 1996 paper in the journal *International Security*, the political scientist Scott Sagan advanced the idea of a domestic politics model of why countries acquire nuclear weapons, where these are used as "political tools" to "advance parochial domestic and bureaucratic interests." (His other two models focus on security threats and prestige as drivers of nuclear weapons programs respectively.)

India offers an excellent example of an institution using nuclear weapons to advance its bureaucratic interests. The Atomic Energy Commission was set up originally in the late 1940s to develop atomic energy for peaceful purposes. The time period was significant for two reasons. Globally, there was an atmosphere of optimistic belief that the power of the atom that had destroyed Hiroshima and Nagasaki could be, and should be, used to generate electricity. Domestically, India had just gained independence in 1947, after two centuries of British colonial rule. The newly independent state's leaders—most importantly, Jawaharlal Nehru, soon to become India's first prime minister—were confronted with enormous development challenges. Foremost among them was to develop the country's productive capacities and building industries, which required increasing electricity generation. Nehru imagined atomic energy providing a perfect solution.

Despite generous budgets and unlimited political support, the Atomic Energy Commission proved inadequate to the engineering task of building nuclear power plants by itself or producing cheap electricity. Until he died in 1964, Nehru never saw a single unit of nuclear electricity being generated. By the mid-1960s, as scholar Itty Abraham reveals in his papers in the *Economic and Political Weekly* and his book *The Making of the Indian Atomic Bomb*, leaders within the nuclear establishment started advancing a different rationale for continued patronage by the state: producing nuclear weapons.

In 1974, that capability was demonstrated to the world through India's first nuclear weapons test. That explosion, and more generally the nuclear weapons program, allowed the nuclear establishment to lay claim to unique institutional status and political power. The source of that power was the twin-promise of perfect security and endless economic growth:

theoretically, all the means for mass production, mass consumption, and mass destruction.[9]

Promise does not equal delivery. To date, nuclear power produces barely 3 percent of the electrical energy in the country. Its contribution is outstripped by wind and solar energy, technologies that were developed only in the last couple of decades. Yet nuclear power continues to be disproportionately patronized by the government, in large part because of the connection to nuclear weapons.

The connection takes a different form in Brazil. Instead of an institution originally set up to develop nuclear energy using military applications of the technology to further its bureaucratic interests, in Brazil the military and its interest in developing nuclear-powered submarines has dominated policy-making on all matters nuclear. The Brazilian military, explain Togzhan Kassenova, Lucas Perez Florentino, and Matias Spektor in a 2020 report, "has become the institutional home to nuclear R&D and a staunch advocate for nuclear power amid budget cuts and criticism from domestic and international actors." Although nuclear reactors produced barely over 2 percent of Brazil's electricity in 2022, these powerful supporters ensure continued patronage for nuclear power.

Corporations and profits

Many large private corporations profit from both these activities and have a vested interest in the acquisition, maintenance, and expansion of nuclear energy and weapons. In the United States, the country that best exemplifies this tendency, the connection was identified in 1961, when outgoing President Dwight Eisenhower cautioned against the "unwarranted influence" and "misplaced power" arising out of the "conjunction of an immense military establishment and a large arms industry" in his farewell address.

Corporations involved in the nuclear weapons enterprise were among the most influential of this arms industry, and many of them also profited from the quest for nuclear energy. Some examples are Westinghouse, General Electric, Babcock & Wilcox (now BWXT), and Bechtel. But as Susi Snyder documents in the 2019 report, "Producing Mass Destruction," many lesser-known companies like AECOM, Fluor, and Jacobs also engage in both pursuits.

AECOM, for example, is a major contractor at the Lawrence Livermore National Laboratory, one of the United States' two nuclear weapon laboratories, and is involved with life extension programs for the B61 nuclear bomb and the W80-1 nuclear warhead. Its website also highlights its role as "engineer or constructor of record" in forty-nine nuclear power plants, including units in Spain, Italy, Brazil, Mexico, and Taiwan, not to mention the United States. Likewise, Fluor has contracts worth billions of dollars for the W88 nuclear warhead, while being heavily invested in the NuScale small modular reactor (see chapter 6). Finally, Jacobs Engineering has a twenty-five-year £25.4 billion contract for maintenance of the UK Trident arsenal, is involved in multiple nuclear decommissioning projects, and has joined hands with Ultra Safe Nuclear Corporation to support the latter's small modular reactor design.

Bechtel

One company that has for decades operated and profited enormously from nuclear energy and nuclear weapons, in the United States and abroad, is Bechtel. According to the analyst and disarmament activist Andrew Lichterman's 2012 report on the company, it has reportedly "provided engineering and construction services at 88% of U.S. nuclear electricity generating plants." It also "received contracts to clean up Three Mile Island in Pennsylvania and Chernobyl in the Ukraine after the disastrous nuclear accidents at those facilities in 1979 and 1986, respectively," reports investigative journalist Sally Denton in her 2016 book *The Profiteers.*

Bechtel also has large contracts for nuclear-weapons-related work. For example, its contract for constructing a chemical plant to deal with the large quantities of highly radioactive wastes at the Hanford site in Washington state, was valued at $12.2 billion in 2006.

And then there is corruption. At Hanford, Bechtel and AECOM have had to pay millions of dollars in fines. In 2016, the two agreed to pay $125 million for knowingly violating "quality standards" and using "substandard materials" in construction projects, and also improperly using "federal funds to lobby Congress to, among other things, try to cut the DOE's budget for independent oversight of work." The second conviction was in 2020 and the two companies agreed to pay $57.5 million in connection with "inflated labor hours being charged to DOE, and for falsely billing DOE for work not actually performed."[10]

Bechtel is also involved in the management or operations of many of the facilities that manufacture, test, and maintain the US nuclear arsenal, including the Los Alamos Laboratory, the Livermore Laboratory, the Pantex plant in Texas (where nuclear bombs and warheads are assembled, refurbished, and dismantled), the Y-12 plant in Tennessee (where the secondaries for thermonuclear weapons are made), and the Bettis and Knolls Atomic Power Laboratories (which provide research and technical support for the navy's nuclear submarines and aircraft carriers). This management structure is referred to as the government-owned and contractor-operated (GOCO) model, which has been very lucrative for Bechtel.

One way Bechtel achieves this status is by cycling its executives into positions within the US government and absorbing government officials who step down. Some Bechtel executives have also served on various high-level government committees wherein they contributed to policymaking.

An example is W. Kenneth Davis, a chemical engineer who headed Bechtel's nuclear division for over twenty years. Davis came to Bechtel in 1958 via the Atomic Energy Commission (AEC), where he was deputy director and the head of the reactor development division. (That could be seen as a trade: another Bechtel executive, John McCone, became AEC chair at the same time.) During the 1960s, Davis, while part of Bechtel, was also the chair of the Committee on Reactor Safety of the Atomic Industrial Forum, where he was involved in setting policies for where nuclear reactors should be sited.

Davis eventually retired from Bechtel to become deputy secretary of energy under President Ronald Reagan. Within the Reagan administration, journalist William Greider revealed in 1982, Davis worked to modify or eliminate the rules put in place under the Gerald Ford and Jimmy Carter administrations to regulate international trade in nuclear technologies that had been devised in response to India's first nuclear weapons test of 1974. Bechtel opposed these rules because they came in the way of profiting from nuclear activities in many countries, especially in the Middle East. Davis was helped by various others within the cabinet, especially President Reagan's secretary of state, George Shultz, another Bechtel person.[11]

Shultz also moved in and out of government. He had already served two stints in US administrations: as secretary of labor and secretary of the treasury for President Richard Nixon. In May 1974, when Nixon's Watergate scandal was at its height and impeachment hearings began, Shultz quietly moved out to become executive vice-president of the Bechtel Corporation, rising eventually to the position of vice-chair of the Bechtel Group, staying in that position till 1982, when President Reagan tapped him. (Well after retiring, Schultz became deeply involved in the now infamous company Theranos, which falsely claimed to have revolutionized the way blood tests were conducted, heavily promoting its CEO Elizabeth Holmes, before she was convicted of fraud.)

With such high-level personnel within the government, it is no wonder that Bechtel has been rewarded with billions of dollars in contracts. But Bechtel did not rely only on such insiders. It also actively lobbied for favorable policies, employing sophisticated manipulation techniques. In the aftermath of the 1979 Three Mile Island accident, Bechtel put an enormous amount of effort into fighting the drastic decline in nuclear reactor orders. Bechtel joined General Electric and a number of electrical utilities to form a lobbying group called the United States Committee for Energy Awareness aimed at maintaining public and government support for nuclear energy. By 1983, the *New York Times* estimated they were spending up to $30 million annually, some of it coming directly from the monthly bills paid by electricity consumers. In 1988, the *New York Times* described the group as "the nuclear industry's main trade association." The committee, explains investigative journalist Sally Denton, placed "supposedly independent energy experts on radio and television talk shows" and submitted "letters to the editors and Op-Eds to dozens of newspapers throughout the country."

Babcock & Wilcox

By the beginning of the twenty-first century, these strategies would lead to creating widespread government support for paper nuclear power plant designs called small modular reactors. The first company to profit massively from that wave of SMR propaganda was Babcock & Wilcox, another company deeply embedded in the military-industrial complex.

Babcock & Wilcox's original claim to fame starting in the mid-nineteenth century was its ability to manufacture steel boilers. During the Manhattan Project, it became a supplier of equipment to the US nuclear weapons program. In the 1950s, B&W went into the nuclear reactor business by supplying one for the first nuclear-powered merchant ship, *Savannah*. Around the same time, it entered the commercial nuclear power plant business, when it obtained the contract for the Indian Point 1 plant in New York. But its best-known project was the Three Mile Island plant in Pennsylvania, which attained notoriety when unit 2 of the plant melted down in 1979.

Babcock & Wilcox never sold another nuclear power plant in the civilian realm. But it managed to maintain a monopoly on supplying nuclear reactors and other components for the nuclear submarines constructed by the US Navy. It is one of the two private firms licensed to process highly enriched uranium and produce fuel for the navy's nuclear submarines.

After the failure of the mPower experiment (see chapter 6) in 2015, Babcock & Wilcox spun off a unit devoted to nuclear power called BWXT, which entered the Canadian nuclear market. BWXT used the flow of money from the navy as its cash base to expand into the nuclear power market, as a supplier not of reactors but of components and services. A July 2017 company profile in *Forbes* magazine gushed, "BWXT not only is the sole provider of naval nuclear reactors, it is also the largest manufacturer of commercial nuclear components in North America, and a key provider of nuclear fuel to civil, military and commercial users." In Canada, BWXT obtained contracts worth hundreds of millions of dollars to supply fuel and nuclear components for ongoing or planned multibillion-dollar nuclear reactor refurbishment projects in the province of Ontario (dressing it up as local economic stimulation in some cases).

Bechtel and B&W are examples of companies that have profited for decades from simultaneous involvement in the military and commercial nuclear sectors. Such companies benefit from many channels for overt or hidden cross-subsidies. Their ability to obtain these subsidies is helped enormously by the ideologically powerful argument that an independent nuclear technology base is in the national interest. Because of these economic factors, corporations involved in these areas continue to support the continued existence of nuclear weapons and the expansion of nuclear energy.

A different beat

After decades of trying to maintain that there was a clear separation between civilian and military nuclear technologies, supporters of nuclear power have been switching the argument in the last couple of decades, especially whenever the nuclear industry is in economic distress. Then nuclear power advocates highlight the close connections and make the case that a healthy nuclear power industry is essential to the production of nuclear materials for war. The pattern is observed in multiple countries. It is remarkable that whenever the nuclear power industry is in trouble, the strongest argument that officials use in order to obtain government support is to emphasize the overlap with military uses. The argument has been particularly potent in the last decade because all the countries with nuclear weapons have been modernizing their arsenals, as detailed in the Assuring Destruction Forever series of reports put out by Reaching Critical Will.

Perhaps the greatest irony is the argument's use in the one country that has suffered from the use of nuclear weapons in war. In 2011, as the Japanese nuclear industry was reeling from the catastrophic Fukushima accident, an official from the Liberal Democratic Party, which has dominated Japanese politics for decades, argued that Japan's capabilities in nuclear power and "leading-edge rocket technology" make it "possible to create nuclear weapons in the relatively short time of several months to a year." This was evidently seen as a strong argument for why Japan should continue to maintain nuclear power, at a time when public opinion had completely turned against the option.

In the United States, the relationship between nuclear energy and military capabilities became, if I may resort to a pun, a trump card during the Trump administration. Given Candidate Trump's characterization of climate change as a hoax made in China, or tweets to that effect, President Trump could obviously not be called upon to support nuclear power for its low-carbon credentials. In chapter 4, I mentioned one reaction from the nuclear industry to his disinterest in dealing with climate change—namely, emphasizing jobs. But another strategy adopted by the nuclear industry and its supporters was to emphasize what they called the "national security" benefits of maintaining the nuclear fleet and subsidizing the industry.

None of the reports and statements produced by nuclear proponents explain how they determined what is national security and why the already oversized military capabilities of the United States had to be maintained and expanded. The primary purpose of that rhetorical thrust was to mark off nuclear power as a national security imperative and prevent any questioning.

The first major report was led by Ernest "Ernie" Moniz, a physicist and nuclear energy enthusiast. Moniz was the founding director of the Massachusetts Institute of Technology's Energy Initiative, overseeing a series of reports on the future of energy technologies. The very first of this series was the 2003 report on the future of nuclear energy, which identified the link between nuclear energy and nuclear weapons proliferation as one of the four unresolved problems associated with the technology.

Fast forward to 2017 when Moniz stepped down from the position of secretary of the Department of Energy during President Obama's second term. In August, Moniz launched a report that called the United States nuclear energy enterprise a "Key National Security Enabler."

Among the many arguments laid out in the report for economic support of nuclear power plants, one explicitly pointed to two material linkages between the supposedly separate industries. First, the "nuclear weapons stockpile requires a constant source of tritium (half-life about 12.5 years), provided by irradiating special fuel rods in one or two power reactors." Second, the report argued that "a strong domestic supply chain is needed to provide for nuclear Navy requirements. This supply chain has an inherent and very strong overlap with the commercial nuclear energy sector and has a strong presence in states with commercial nuclear power plants."

Also in 2017, the Nuclear Energy Institute, the advocacy group for the nuclear industry, lobbied Congress to pass legislation that would extend tax credits to nuclear power plants under construction. Among their arguments was the claim that if these plants were not constructed, then "it would also stunt development of the nation's defense nuclear complex, because the engineering expertise on the energy side helps the defense side." In other words, trained people possess expertise in cutting across these compartments.

Many of these talking points were echoed the following year in a leaked May 2018 Department of Energy memo:

> The entire US nuclear enterprise—weapons, naval propulsion … depends on a robust civilian nuclear industry … a significant portion of our naval fleet relies on nuclear power. The Navy has over 100 nuclear reactors in ships and submarines, and if civilian capabilities were to deteriorate further, U.S. nuclear defense capabilities (infrastructure, supply chain and expertise) will similarly suffer. Importantly, the civil nuclear industry supports the Navy as a synergistic partner for personnel and supply chain. University nuclear engineering programs supply both the nuclear navy and civil nuclear industry with highly trained personnel, and the civil nuclear industry provides an attractive employment opportunity following military service. Absent a vibrant civilian industry, university programs contract or collapse. The civil nuclear industry helps support the supply chain of over 700 companies in 44 states, which are also relied upon by the nuclear navy.[12]

The chorus grew louder the following month, when "several dozen retired generals and admirals, former State, Defense and Energy Department officials, three former chairmen of the Nuclear Regulatory Commission, and a sprinkling of former senators, governors, industrialists" wrote a "letter to Energy Secretary Rick Perry attesting to the connection between U.S. nuclear power plants and national security." The letter, which is hosted on the website of the Nuclear Energy Institute at the time of this writing, asserts that the "national security benefits of a strong domestic nuclear energy sector take many forms, many of which overlap" and highlights, among other points, the fact that "many of the companies that serve the civil nuclear sector also supply the nuclear Navy and major DOE programs."

By this time, anyone who was within the Washington Beltway must have received the message. But for good measure, the Atlantic Council, another Washington-area think tank, set up something called the Nuclear Energy and National Security Coalition, "in order to increase awareness about nuclear energy as a cornerstone of national security." The council selected a panel of luminaries to be part of the task force, including a veritable who's who of the US civilian and military nuclear programs.[13] To be seen as nonpartisan, the group chose two US senators, one Republican and one Democrat, as co-chairs. The following year, to no one's surprise, the group's report asserted: "The US civilian nuclear power industry is a strategic asset of vital importance to US national security."

The academy, another sector that often benefits from the US government's interest in nuclear technology, added its heft too. In a 2018 article in the *Electricity Journal*, three academics from the University of Georgia at Athens harkened back to Cold War history to contend that nuclear energy is "capable of reshaping the geopolitical contours of the world order." These academics then went on to call for the United States to be "engaged as global experts in the international control of the nuclear supply chain."

This was certainly the case during the Cold War. International relations scholar Lauren Richardson has argued that US nuclear power companies "viewed South Korea as an attractive business prospect" because they "had a specific agenda to promote the advancement of nuclear technology in noncommunist countries." Likewise, Russia and China also use nuclear power for geostrategic purposes, as nonmilitary means of influencing other countries.

There was a parallel effort in the United Kingdom to drum up support for nuclear energy by emphasizing the overlap with the military. British academics Phil Johnstone and Andy Stirling have documented how these interlinkages between the nuclear power and weapons programs started being publicly emphasized once the United Kingdom began grappling with the complexities of constructing the Hinkley Point nuclear power plant and analogous problems with building a new Trident nuclear-powered submarine.

During the 2009 hearings of the Parliament House of Commons Innovation, Universities, Science and Skills Committee, the Institution of Nuclear Engineers and the British Nuclear Energy Society highlighted how the nuclear submarine program serves to train nuclear power plant personnel. The Dalton Nuclear Institute at the University of Manchester (see chapter 4 for another of its talking points) explained that the opposite is also true, that "reactor physicists" can "develop their skills and knowledge by researching civil systems" and "when necessary" use these for developing military systems. All this came with the warning that these linkages "need to be carefully managed to avoid the perception that civil and military nuclear programmes are one and the same." Evidently, the aim is to ensure that the unpopularity of one activity does not affect the other.

The premier nuclear company in the United Kingdom is Rolls Royce. Once known the world over as a manufacturer of luxury cars before it sold this division to BMW, Rolls Royce now gets most of its revenues (over 70 percent in 2021) from civil aviation and defense, but is increasingly positioning itself as a sustainable power company. As part of this reinvention, it is developing a nuclear reactor design called the UK SMR for which it sought and received funding from the British government.

To advocate for government funding, Rolls Royce's 2017 introduction to its small modular reactor design promoted precisely the linkage between civilian and military nuclear technology, arguing that a government program on these reactor designs would "assist in sustaining the skills required for the Royal Navy's submarine programme." Rolls Royce offers an economic motivation by arguing that expanding "a nuclear-capable skilled workforce through a civil nuclear UK SMR programme would relieve the Ministry of Defence of the burden of developing and retaining skills and capability. This would free up valuable resources for other investments."

The attraction of nuclear power to the military industry has also been on display in Australia. So far, the country possesses neither nuclear power plants nor any military nuclear technology (except indirectly through its defence relationship with the United States). Yet when the Australian government announced plans to build new conventionally powered submarines, the Submarine Institute of Australia went on to canvass for including nuclear-powered submarines. The institute also went on to host two seminars in 2021 emphasizing the nexus between developing nuclear power plants for electricity generation and nuclear submarines.

One final example comes from India. Its Department of Atomic Energy (DAE) publicly emphasized the connection between its nuclear power and nuclear weapons programs during the negotiations in the aughts over a political deal with the United States. One key disagreement during those negotiations was over how many nuclear power plants would be opened up to the International Atomic Energy Agency, to verify that plutonium produced in these reactors was not being used to make nuclear weapons.

US negotiators and their counterparts in India's Ministry of External Affairs were pushing for a sizeable number of these reactors to be put under safeguards. The DAE pushed for putting as few as possible under safeguards. Its strategy was to designate even civilian nuclear power reactors as military ones. While this debate was being carried out behind closed doors, the head of the DAE, Anil Kakodkar, decided to go public during a 2006 interview with senior journalist Pallava Bagla.

One question posed by Bagla was "Is your strategic need for plutonium not met by CIRUS and Dhruva [two reactors already producing plutonium for nuclear weapons]? Do you need additional capacity from civilian reactors?" Kakodkar replied in the affirmative but with a significant correction: "Yes, very clearly. Not from civilian reactors, but from power reactors." The implication of the correction is that, for the DAE, a reactor's use for making nuclear weapons material is not to be precluded just because its main purpose may be to produce electricity.

Kakodkar's statement was an unprecedented public statement on a policy under negotiation. Strictly speaking, it would be a violation of India's conduct rules for government officials. The fact that Kakodkar intervened in this fashion is testimony to the DAE's political power and ability to shape decision making, and that derives from the fact that the agency is involved in the pursuit of both nuclear energy and weapons.

At a personal level, I can also testify to the power of this argument. On many occasions, my critiques of nuclear energy in India—its marginal relevance to the country's energy needs, the high costs of generation, the environmental impacts, the safety risks—are dismissed with a statement along the lines of "But it provides us the bomb and all the power and prestige that come with that." This, from otherwise thoughtful and intelligent people.

Relevance

Understanding the specific connections between nuclear weapons and nuclear energy is essential to comprehending why governments continue to support nuclear power despite the myriad problems associated with the technology. As this chapter detailed, the linkages between nuclear weapons and nuclear energy benefit organizations involved in either pursuit, or both. Public sector organizations and private corporations use these connections to control information flows and exclude outsiders from policy discussions, thus weakening democracy. The powerful ideological appeal of nuclear weapons adds to the problem.

The expansion of nuclear energy also thwarts efforts toward a world free of nuclear weapons. It will not be possible to eliminate nuclear weapons without policies and resource-allocation decisions that are grounded in the reality that nuclear energy cannot be separated from nuclear weapons.

Conversely, the existence of nuclear weapons anywhere in the world allows corporations and countries to continue to call for the investment of public resources into uneconomical and environmentally damaging nuclear power plants and other nuclear facilities. The dangers associated with, and the impacts of, nuclear weapons and nuclear energy might be quite different, but attempts to grapple with them cannot be compartmentalized; they have to be confronted simultaneously.

Elsewhere in the book, I have argued that stopping the expansion of nuclear energy would help with climate mitigation. Here, I end with a quote from Ted Taylor, the weapons designer I began this chapter with, that elucidates what else can be accelerated with such a step: "A complete phaseout of nuclear power would help focus the world's attention on safeguarding nuclear materials and safe, permanent disposal of all the nuclear wastes and spent nuclear fuel, separated plutonium, or other stockpiles of nuclear weapon materials that had been produced before nuclear power is completely phased out."[14]

6

Magical Thinking and Billionaire Messiahs: New Technology as Solution?

Right now it's only a notion. But I think I can get money to make it into a concept, and then later turn it into an idea.

Annie Hall, 1977

Those most responsible for creating the problem [of climate change] will see to it that they profit from the solution that they propose.

Arundhati Roy, 2019

One slightly cloudy morning last year, as I raced to meet my editor's deadline for this manuscript, I received a surprising email: "In the fierce competition to bring Small Modular Reactors to the UK, the UK/Ireland Nuclear Free Local Authorities have recently noted that one new applicant to the market stands out. Like its Rolls Royce rival, this design would be prefabricated and its components would also be assembled on site, but, unlike any conventional nuclear power station or modular reactor, it will have a more affordable price tag, will not take years to build or go over budget, comes without any environmental contamination or radioactive waste, is not dependent on uranium sourced from any nation led by a despot, and cannot become a dirty bomb if attacked by a hostile power in any future conflict involving this nation."

The reactor design's description is remarkably similar to a number of nuclear reactor designs promoted in recent years by billionaires and the media. I found these claims surprising, because the email came from a group wanting to be free of nuclear weapons and energy. Only when I went to the accompanying link did I realize what the date was: April 1, 2023. An April Fools' joke. Duh!

Maybe I should have been smarter. But I do see stories like this all the time in my email inboxes. And even though I fell for an April Fools' joke, I don't feel much like a fool, given how many times the nuclear industry successfully has sold a questionable technology as the solution to our energy problems, and the number of times some acquaintance has forwarded a breathless news article about a breakthrough nuclear reactor technology.

One example is Transatomic Power. About a decade ago, *Technology Review,* a magazine that comes out of the Massachusetts Institute of Technology, ran an article titled "What If We Could Build a Nuclear Reactor That Costs Half as Much, Consumes Nuclear Waste, and Will Never Melt Down?" The article was a bio-feature of one of the founders of Transatomic, Leslie Dewan, then a graduate student at MIT.

Dewan and her colleagues had started with an old concept called a molten salt reactor and made what they described as "some relatively modest changes." The result was a design that was, according to the designers, "much more power dense" and "a lot cheaper." The reactor, Dewan claimed, "is now so compact that a version large enough for a power plant can be built in a factory and shipped by rail to a plant site, which is potentially cheaper than the current practice of building nuclear reactors on site." If all of that was not enticing enough, Dewan explained, the reactor "can consume about one ton of nuclear waste a year, leaving just four kilograms behind."

Later in this chapter, I will briefly describe the problems with such designs. For the present, let us continue with the Transatomic story.

Promises of the sort listed above proved appealing to a bunch of investors. In 2014, Founders Fund—a San Francisco–based venture capitalist company whose best-known partner is Silicon Valley billionaire Peter Thiel—chipped in $2 million as seed funding. The following year, Transatomic raised an additional $2.5 million from Founders Fund, Acadia Woods Partners, and Daniel Aegerter, chair of the Swiss fund Armada Investment AG. In 2016, the US Department of Energy also awarded $200,000 to Transatomic.

Meanwhile, the founders of Transatomic and their employees advertised widely their design's safety and capacity to reduce the amount of waste. A December 2015 *Popular Science* article reporting on Thiel's investment and enjoining others to "go nuclear" also quotes Dewan describing Transatomic's reactor design as something that can "simmer like a Crock-Pot for decades." In other words, she portrayed the reactor as not just safe but domesticated, something that can be brought into one's kitchen. And a flashy documentary called *The New Fire* prominently featured Dewan and the co-founder of Transatomic Mark Massie explaining how they hope to solve the climate problem by building nuclear reactors everywhere.

Transatomic's greatest asset, arguably, was an impressive list of supporters. Advising it were MIT professors (Richard Lester, Benoit Forget, and George Apostolakis, who had also served on the Nuclear Regulatory Commission), senior nuclear bureaucrats (Jess Gehlin, then at the Oak Ridge National Laboratory; Todd Allen, then at Idaho National Laboratory; Michael Corradini, then the president of the American Nuclear Society), and senior officials from the nuclear industry (Regis Matzie from Westinghouse). To top it all, in 2017, Oak Ridge, the Department of Energy laboratory with experience operating a molten salt reactor, albeit back in the 1960s, announced that its scientists had "verified the viability" of the Transatomic design to reduce nuclear waste substantially and described the reactor technology as "walk-away safe."

With such luminaries reviewing Transatomic's ideas, there was no reason to disbelieve any of their claims. Or was there?

It turned out that in 2015, Kord Smith, an MIT professor who was not part of this happy group, had discovered serious flaws in the reactor's design. As Smith told *Technology Review* in February 2017, Transatomic's assertion about how much electricity it could generate from a given amount of uranium was "obviously incorrect based on basic physics." Because he was also at MIT, Smith could ask the company to run a test—something outside critics could not demand. The test confirmed Smith's hunch: Transatomic's claims about the efficiency of their reactor proved to be "completely untrue."

So much for Oak Ridge's scientists or Transatomic's other highly qualified advisors.

The problem Smith identified was serious and turned out to be the beginning of the end for Transatomic. In September 2018, Leslie Dewan posted the final note on the Transatomic website, where she confessed, "Despite our best efforts, we haven't been able to scale up the company rapidly enough to build our reactor in a reasonable time frame. It is therefore with a heavy heart that I must announce that Transatomic will suspend operations."

Transatomic's story encompasses both typical and unusual elements. What was unusual about Transatomic was its having to publicly backtrack on false assertions. But its tall claims are typical of the hype nuclear reactor designers regularly put out; its success in raising funding from private investors is also typical in a time where there is easy money to be had by anyone claiming to be doing something innovative. After all, private investors have to find companies to invest in—that is what defines them as investors. The debacle at Transatomic, for example, did not prevent Founders Fund from investing in another nuclear startup—Radiant—which was developing a high-temperature gas-cooled reactor called Kaleidos.

In this chapter, I start by briefly describing the investment landscape for new nuclear reactors. I then explain why such designs will not be economically competitive, and argue that no reactor design can solve the multiple problems confronting nuclear energy, including high costs, safety, proliferation, and waste.[1] I also outline the history of nonstandard nuclear technologies, including high-temperature gas-cooled reactors (the type Radiant wants to build), sodium-cooled fast neutron reactors (what Bill Gates wants to build), and molten salt reactors (what Transatomic wanted to build). Finally, I examine the tactics adopted by these investors to profit from their investments, ending with a brief outline of the ideological role that nuclear power plays in the minds of these investors, and in the future they envision.

Billionaires for nuclear energy

Transatomic is by no means the only nuclear design company that investors put money into. Over the last decade and a half, after the financial crash of 2008, a number of investors have provided capital for nuclear energy companies. By 2015, Third Way, a pro-nuclear think tank, compiled a list of over forty-five companies that had received a total of $1.3 billion in private funding to develop reactors based on either nuclear fission or fusion. The first name in the list was Transatomic Power.

Although the amounts invested are relatively small when compared with the historical costs of developing and testing nuclear reactor designs and having them licensed by regulatory authorities, nuclear advocates marshalled such statistics to create the impression that the nuclear sector is growing and ripe for more investment. The following year, Third Way partnered with various national laboratories to host what it billed as "a first-of-its-kind Advanced Nuclear Summit and Showcase in Washington" featuring senior members of the US Senate. The images that Third Way generated made their way to a Strategic Vision document published by the Department of Energy's Office of Nuclear Energy in 2021.

The buildup was intense. News media also highlighted the private sector investing in nuclear reactor startups. In 2014, Harvard Business School produced a case on NuScale Power with the subtitle "The Future of Small Modular Reactors," which envisioned a couple of hundred SMRs being built during the 2020–35 time frame. The hype and the investment fed into each other.

Many of the people behind such investments are high-profile individuals. Well-known names frequently featured in the media are Bill Gates and Peter Thiel. Other prominent billionaires investing in, or simply promoting, nuclear power include Sam Altman and Elon Musk.

Although very different in their public profiles, both Gates and Thiel are concerned about climate change. Thiel has gone very far in planning for a climate-related doomsday—literally as far as purchasing a mansion in New Zealand because he expects it to be safer there in the event of an apocalypse, reveals a September 2018 article in *Bloomberg*.

Bill Gates, for his part, wrote a book explaining "How to Avoid a Climate Disaster" and promoted it extensively. In a subsequent interview to CNBC in February 2021, Bill Gates announced: "There's a new generation of nuclear power that solves the economics, which has been the big, big problem."

Gates's financial contribution to this "new generation of nuclear power" has been through a company called TerraPower. Founded in 2006, TerraPower has featured Gates as the chair of the board continuously, at least as of 2024. In June 2010, the *New York Times* reported that the company received $35 million in seed money from venture capital firms to develop the first of its nuclear power plant designs, the "traveling wave" reactor. Breakthrough Energy Ventures, an investment firm cofounded and cochaired by Gates, has also provided an undisclosed amount of funding.[2] Gates also promised to invest $1 billion from his personal coffers— and raise another $1 billion in private capital—to fund TerraPower directly.

Despite these announcements, how much Gates has person-ally invested is not publicly known. When the *Washington Post* requested an interview about his investment in the company, Gates declined. TerraPower's financial records are not publicly available.

Secrecy notwithstanding, these figures make for excellent media coverage. Even a few million dollars of "seed" funding is sufficient to hire a handful of nuclear engineers—including graduate students—and buy some fancy computers and spe-cialized software. Because most of these supposedly innovative advanced reactor designs are just reruns of designs first studied in the 1940s to the 1960s, these investments suffice for pro-ducing a rough conceptual design, the stage just a bit more advanced than what is described in the *Annie Hall* quote at the beginning of this chapter (i.e., a concept).

Additionally, a lot of the initial research and development is done on the taxpayer's dime. The NuScale reactor design, for example, was the outcome of the Multi-Application Small Light Water Reactor project funded by the US Department of Energy and largely carried out in two public institutions: Idaho National Laboratory and Oregon State University.[3] Likewise, the virtual reality tools used by companies like Westinghouse (described later) were funded by the DOE, and carried out at another public university, The Pennsylvania State University.

Going from that early stage to a design detailed enough to convince a safety regulator about the low probability of a serious accident is much harder. Any good regulator will pose many tough questions. Answering them will not be easy, requiring "several million person-hours of design/engineering work" to reach "the level of technical confidence demanded by regulatory authorities" according to a 2018 report by a group of MIT nuclear engineers. (As mentioned in chapter 2, it is expected that the EPR2 reactor design will require 20 million person-hours to complete.)

All that labor is reflected in the price tag. A 2015 US Government Accountability Office report estimated that developing a new nuclear reactor design and obtaining the US Nuclear Regulatory Commission's certification "can cost up to $1 billion to $2 billion," a figure corroborated by the 2018 MIT report. That might be an underestimate, going by the case of NuScale, the leading SMR vendor in the United States. In a November 2023 earnings call, the NuScale CEO declared that the company had invested more than $1.8 billion. But the reactor hasn't been licensed and the regulator still has many questions about the design that NuScale hopes to build.[4]

That is a lot of money, and most investors don't commit anywhere close to those amounts—even those that are personally wealthy enough to write a check for a billion dollars without missing it. Instead, after their initial investment, they rely on the government to pick up a substantial part of the tab. For example, TerraPower, the company backed by Bill Gates, received a $40 million grant in 2016, followed by another $80 million in 2020, and $8.5 million in 2022, all from the Department of Energy. Further, the 2021 Infrastructure Investment and Jobs Act has earmarked $2.5 billion for nuclear projects, and a TerraPower nuclear project proposed for the state of Wyoming is expected to receive part of this funding.

For TerraPower, government support might add up to nearly as much as private investments, at least as far as the publicly available information can tell us. Bill Gates also paid numerous visits to China and nearly succeeded in getting the people of that country to invest their RMBs into his idea. Unfortunately for Gates, but luckily for the Chinese, that idea was stopped by America's waning diplomatic and trade relationship with China.

TerraPower is not alone in seeking public funding. Just about any company working on new nuclear reactor designs, whether it is Transatomic or NuScale, has received large amounts of taxpayer money. A paper published in *Environmental Research Letters* in 2017 calculated that between 1998 and 2015, companies and institutions working on "advanced nuclear" reactors received about $2 billion in US government funding. All for nothing. Not one of the projects that received any part of the $2 billion saw the light of day.

Learning nothing from these failed ventures to develop "advanced nuclear reactors," the US DOE also funneled millions of dollars into small modular reactors. The budget watchdog organization Taxpayers for Common Sense has calculated that between 2011 and 2021, the DOE has spent "more than $1.2 billion on SMRs" and has announced further awards over the next decade that could amount to "at least $5.5 billion more" than what has already been awarded.

Nor do these investors want to wait for the couple of decades it will take for the nuclear reactor concept to be fully developed, scrutinized by a regulator, and power plants sold.

Instead, they push the company to "go public"—namely, sell stock to the public as early as possible. Ray Rothrock, a venture capitalist and investor, was asked by a panel of the US National Academies of Sciences, Engineering, and Medicine in January 2021 how investing in such nuclear reactors makes money for people like him. Rothrock's answer, in essence, was that investors make a lot of money when the company goes public, and this happens well before these reactors sell any energy.

This is, in essence, a standard Silicon Valley strategy. The primary purpose of the new reactor design is not to produce energy safely or cheaply, but to make money for these investors in the short term. Such investments also serve an implicit ideological function: convincing others that nuclear ventures will be profitable. Hype begets hype.

In recent years, there has been a tendency for nuclear companies to go public by setting up what is called a special purpose acquisition company (SPAC). At least three SMR companies have chosen this route: NuScale, X-energy (which subsequently abandoned this plan), and Oklo. In July 2023, Oklo announced plans for a merger with a SPAC called AltC Acquisition

Corporation, which was cofounded by Sam Altman, the CEO of the artificial intelligence company OpenAI; Altman is also the chair of Oklo's board.

SPACs, explains a February 2021 article in *Harvard Business Review*, are "shell companies that have no operations or business plan other than to acquire a private company using the money raised through an IPO (Initial Public Offering), thereby enabling the latter to go public quickly." Because the company goes through an IPO with no prior business activity, it offers no records to scrutinize its actions. Given all the problems with nuclear energy, one can see why companies developing new reactor designs might be attracted to this option.

But then, when the shell company merges with another company, the latter effectively becomes a public company, without having undergone checks meant to protect investors from scams and excessive risk. Even the US Securities and Exchange Commission has been concerned about the growth in SPAC transactions. Nevertheless, going public keeps the hype cycle up.

Fortunately, the market can sometimes discern the obvious problems with SMRs, and going public does not always result in an influx of cash, as we shall see in the case of NuScale. And the developers of the mPower design found out even earlier that there was no demand for their product.

No path to commercial success

The mPower design was developed by Babcock & Wilcox, a veteran of the nuclear business (more on this company in chapter 5). So, it was not surprising that it was the first company selected to receive up to $226 million under the 2012 DOE cost-share funding opportunity to "guide two small modular reactor designs through the nuclear regulatory process by 2022."

For a little while, mPower seemed poised to make history. When US DOE selected mPower, James Ferland, president of Babcock & Wilcox, pronounced that the award represented "another key milestone in the work to establish *the world's first commercially viable SMR nuclear plant*" (my emphasis). In 2013, the Tennessee Valley Authority, one of the country's large utilities, signed a contract geared toward "deploying up to four mPower SMR reactors." Ferland announced that the goal was "completing the first commercial deployment of an SMR in the United States by 2022." The *New York Times* described the company as being "in the lead" in the race to develop SMRs, in part because it had the Energy Department and the Tennessee Valley Authority "in its camp."

The lead turned out to be short-lived, and the outlook dimmed once Babcock & Wilcox started looking for other investors or customers. Despite the hype, few were willing to invest in an mPower reactor or buy the expensive power it might generate at some future date. By 2014, Babcock & Wilcox had to admit that neither was forthcoming. The company drastically reduced the budget for mPower. A subsequent attempt to involve a partner firm, Bechtel Corporation, also ended in failure, and by 2017, Babcock & Wilcox had essentially abandoned the mPower project. According to nuclear enthusiast Rod Adams, Babcock & Wilcox spent "about $400 million" on the failed mPower project.

Compared with Transatomic, Babcock & Wilcox was lucky and could afford to abandon this project because it had many other ways to make money. The other company funded by DOE as part of its 2012 cost-share funding opportunity, NuScale, did not have such options. So, even though its business prospects are no better, NuScale has continued operations. In 2015, NuScale finally managed to reach an agreement with the Utah Associated Municipal Power Systems (UAMPS), a collection of small publicly owned municipal utilities in Utah, California, Idaho, Nevada, New Mexico, and Wyoming, to set up a power plant. Again, its proponents posited that it "could be the first SMR project in the world" and that it "provides the safest, most cost-efficient approach."

Reality proved a harsh judge of such promises. Those early claims about cost efficiency proved completely premature. Indeed, as of 2023, NuScale has not managed to even come up with a stable reactor design, and the company had to resort to successive power increases, from an original forty megawatts to seventy-seven megawatts.[5] Although NuScale is unlikely to admit it, these increases testify to diseconomies of scale (discussed in the next section) and the folly of developing a small reactor in the first place.

Like the cost of the AP1000 or many other reactor designs before, NuScale's estimated cost has continuously increased. In February 2018, UAMPS estimated that the then 720-megawatt project would cost about $4.2 billion. By 2020, that cost estimate had increased to $6.1 billion. The cost increase drove a number of UAMPS members to pull out of the project.

By January 2023, the cost estimate went up even further, to an eye-popping $9.3 billion for just 462 megawatts of power capacity. In per-kilowatt terms, that estimate for the UAMPS project is around 250 percent more than the initial per-kilowatt cost for the Vogtle project in Georgia. Finally, in November 2023, UAMPS and NuScale terminated the project because of insufficient interest. In January 2024, NuScale laid off about 30 percent of its full-time staff.

Back in 2008, when NuScale was incorporated as a company, its leading officials announced that one of its plants "could be producing electricity by 2015–16." In January 2023, NuScale expected that the first reactor will start producing power in 2029. With the cancellation of the UAMPS project, even that elongated timeline will not be met.

NuScale also went public using the SPAC route by merging with Spring Valley Acquisition. But the market did not react in the way investors expected. Prospects for growth seemed bad enough that the company's chief financial officer and chief technical officer and other insiders sold $4.5 million worth of stocks in the summer of 2023. NuScale's stock price declined from a maximum of $14.87 in August 2022 to a low of $1.92 in January 2024.

Problems with SMRs

The public ought to be skeptical of investing in SMRs. As their name suggests, small modular reactors produce relatively small amounts of electricity—less than 300 megawatts by definition—in comparison with the reactors that have been built over the last half century. When the power output of the reactor decreases, it generates less revenue for the owner, but the cost of constructing and operating the reactor is not proportionately less. All else being equal, a (large) reactor that produces, say, five times as much power as a small modular reactor does not need five times as much steel or five times as many workers. These so-called economies of scale are often touted by the nuclear industry: when South Carolina Electric & Gas Company made its case for building two AP1000 reactors, it pointed to Westinghouse developing the AP1000 design by applying "economies of scale to the AP600 design to reduce the cost per kW." Economies of scale are not limited to nuclear reactors: a bus is a cheaper way of transporting a hundred people than twenty cars.

SMRs, on the other hand, will suffer from diseconomies of scale. They will have greater material and labor requirements relative to their power output, and thus be more expensive than large ones when these costs are weighted by electricity-generation capacity. A hypothetical example might help clarify. Imagine that a country or a utility company anticipates a shortfall of 1,000 megawatts of power sometime in the future, and it has decided to build nuclear reactors to meet this projected demand. If its choices are a large reactor designed to generate 1,000 megawatts of power and a small reactor designed to generate 200 megawatts of power, meeting the demand for 1,000 megawatts of power would mean building either five small plants or one large plant. Because they don't benefit from economies of scale, the cost of the 200-megawatt plant might be about 2.5 times less, not five times less, than the 1,000-megawatt plant.[6] Then, the cost of building five small reactors would be twice the cost of one large plant. Operating and fueling these five reactors will also be costlier than the corresponding cost for the larger reactor. The result will be more expensive electricity from smaller reactors.

Below I discuss how this affected the earlier set of small reactors that were built in the United States. But for the present, let us focus on the current narrative offered by SMR proponents.

SMR proponents argue that there are ways to save money during manufacture and these savings will compensate for diseconomies of scale. These savings, according to this fictional narrative, comes from manufacturing SMRs en masse in factories —hence the term "modular" in small modular reactors— and by learning from the experience of building many reactors.

Let us start with factory manufacture. Tony Roulstone, a nuclear engineer who has worked at the UK Atomic Energy Administration and Rolls Royce for decades, explains that the key idea is "to transfer much of the complex construction work from site to factory conditions" because "productivity is much higher" in factories.

We have heard such stories. When it started marketing the AP1000 and AP600 reactors, Westinghouse promised that it would reduce cost and the time taken to build these reactors by utilizing "modular construction techniques." Wielding language virtually identical to contemporary SMR descriptions, Westinghouse emphasized how its design maximized the use of modules, which would be built in factories and shipped to the plant site. Jill Clelland, a lead manager at

Westinghouse, promised that the reactor could be built in "36 months."

Reality proved otherwise; the AP1000 reactors built in the state of Georgia took more than thrice as long. Factory construction did not avoid all the problems plaguing building reactors at the final site (see chapter 3). But there were also problems in the factory. There is no reason to expect SMRs or other reactor designs to be problem free.

The second claim, about the cost of reactors decreasing by learning from repeated construction, does not hold up to scrutiny either. To start with, the cost may not decline even with many reactors being built. As discussed in chapter 2, reactor construction costs rose, not fell, as utilities built more reactors. The increase partly resulted from the financial costs of addressing newly discovered accident pathways. One can expect similar increases with newer reactor designs too.

Second, smaller plants have a higher cost per unit of power capacity. Even if one were to make very optimistic, and historically untenable, assumptions about cost declines due to learning, smaller reactors would have to be manufactured by the hundreds if not the thousands to break even with large reactors.[7] But large reactors themselves cannot compete economically against renewable alternatives. Ergo, SMRs will never rival renewables in cost.

Finally, even if all these unlikely circumstances materialized, mass-manufactured reactors would give cause for worry. Energy analyst Arjun Makhijani has highlighted the problem of recalls, the underbelly of mass manufacturing, as he put it. Should there be a safety problem with any SMR, all the other SMRs fabricated in the same factory, or even all SMRs using the same design or the same flawed component, might have that problem too. Manufacturers would have to recall all these SMRs, just as Boeing had to recall its 737 Max airplanes and Toyota its Corolla cars. Transporting a radioactive reactor back to the factory is no simple matter. Figuring out what to do about an electricity system relying on factory-made identical reactors under recall is an even bigger headache.[8]

History repeating itself?

The essential idea of using small nuclear reactors to produce energy in a variety of contexts dates back to the late 1940s, when nuclear energy was first discovered and extolled.[9] Well before anyone started building commercial power plants, the United States Army, Air Force, and Navy each initiated research and development programs for various types of small nuclear reactors. For the air force and navy, nuclear energy offered a new way to power their preferred vehicles for delivering death and destruction: long-range bombers in the case of the air force, and submarines and aircraft carriers in the case of the navy. In 1961, the air force program was terminated, after it had spent more than $1 billion; as President Kennedy explained, "achieving a militarily useful aircraft" was not possible in the foreseeable future.

The navy saw more success, at least technically: it did produce nuclear-powered ships and submarines. But these failed the test of the marketplace, as shown by the case of the *Savannah*, the first nuclear-powered ship. During its launch in 1959, officials described the *Savannah* as "potentially the greatest revolution in maritime history." But it quickly became apparent that it cost too much to operate. By 1971, operating costs outpaced revenues by $2 million, and the ship stopped operating. *Savannah* today has been relegated to storage "in a remote corner of Baltimore Harbor" according to an April 2023 *National Geographic* article.

The closest analogs to today's SMRs come from the Army Nuclear Power Program: eight small reactors, some of which were built in isolated locations like Antarctica and Greenland. These reactors, too, proved problematic, and the army shut down all of them. The PM-3A reactor at McMurdo Sound in Antarctica, for example, developed "several malfunctions, including leaks in its primary system [and] cracks in the containment vessel," according to the official history of the army's nuclear power program. Leaks from the plant resulted in soil being contaminated. The army ultimately had to remove large quantities of this soil and ship the contaminated consignment to Port Hueneme, a naval base north of Los Angeles, for disposal.

Even as these military-funded efforts spluttered, the US Atomic Energy Commission funded the construction of several small power reactors, which were, at least in the eyes of the commission, suitable for use in rural areas and for foreign export. None of the reactors funded by the Atomic Energy Commission operated till their promised lifetime.

Illustrative of the economic challenges faced by small nuclear reactors is the twenty-two-megawatt Elk River plant built about fifty kilometers northwest of Minneapolis, Minnesota. A December 1956 advertisement from its operator dubbed it "Rural America's First Atomic Power Plant." As with SMRs proposed today, Elk River used prefabricated components, and transported components to the site using a standard railroad flat car. But all those practices didn't save the project.

Cracks appeared in the cooling system piping of the reactor within four years after it started. Faced with high repair costs, the operating company decided to shutter the reactor. Its spokesperson told the *Chicago Tribune* in December 1971 that it didn't want "to spend the money, especially since the reactor has not been too economical because it is too small," adding that the reactor had produced power at twice the cost of power from coal-fired plants.

This case provides an early example of what we discussed above: diseconomies of scale. The trend continues—several such units have been permanently shut down in recent years. Even nuclear power enthusiasts acknowledge that smaller nuclear plants "tend to be unprofitable more often than do large ones."

Yet nuclear advocates continue to make tall claims about the costs of unbuilt nuclear reactor designs. That is easy. It is also easy to be hopelessly wrong. Shortly after the first commercial nuclear power plant in the United States at Shippingport, Pennsylvania, started functioning, John Gray, the manager of the team that built the plant, presented their experience with that plant at the 1958 Atomic Industrial Forum meeting. As recalled in the 1994 book *Nuclear Power from Underseas to Outer Space*, when Gray reported the estimated cost of power as 64.4 mills per kilowatt-hour (or 6.4 cents per kilowatt-hour), one questioner asked: "Mr. Gray, why is your cost so high when those planning new plants predict they will produce power at 5 to 10 mills per kilowatt hour?"

Gray's reply to that question remains relevant: "Shippingport was built and we kept books." Shippingport produced sixty megawatts of power, roughly the same as the seventy-seven megawatts a NuScale module would produce. If and when new small modular reactors are built, and if books are kept, they will show that electricity from these will still be far costlier than what nuclear advocates claim today.

The false promise of advanced reactors

When such problems are highlighted, nuclear proponents offer another argument. These problems belong to the past and are relevant only to older reactor designs. Newer reactor designs, which go by names like "advanced" reactors, or "generation-IV" reactors, or some other such fancy name, will not be afflicted by these issues.

The term "advanced" is arguably the most sweeping of these. If one goes to their historical origins, even the large reactor designs being offered by US nuclear reactor companies can be termed advanced. That's because these reactor designs all benefitted from the US Department of Energy's Advanced Light Water Reactor Program initiated in the 1980s. That program aimed to "restore nuclear power as a viable option" in the 1990s, by producing "improved and simplified" reactor designs that were to be "economically attractive, and constructable on a predictable and reasonable schedule," an official from the DOE told the 1990 American Power Conference. This program underwrote the AP1000 reactors that proved disastrous to the consumers of Georgia and South Carolina (see chapter 3).

These days nuclear advocates shift the definition and use terms like "advanced" to refer to reactor designs not cooled by water. That cooling role is played by gases like helium, or molten metals like sodium, or by molten chemical compounds called salts (of which common salt—sodium chloride—is an example). Many of these reactor designs also fit into the category of small modular reactors because they would produce less than 300 megawatts of electricity.

But earlier generations of nuclear designers have explored just about all these types of reactors—and abandoned them because of technical or operational problems. It is impossible to go through each reactor design, but I will explain some of the problems that afflict three generic categories of reactors.

High-temperature gas-cooled reactors

Perhaps the most prominent example of this variety of reactor is designed by X-energy, which describes itself as a "leading developer" of SMRs, offering "a Generation-IV reactor technology with a proven operational pedigree." The US government will likely agree with the first part of that statement, since two wealthy and powerful government institutions have funded X-energy: the Department of Energy and the Department of Defense. The other description—"proven operational pedigree" —is true. But what has been proven?

The idea of high-temperature gas-cooled reactors dates back to 1944, even before the bombing of Hiroshima and Nagasaki, when Farrington Daniels proposed a "high temperature pebble pile" to produce plutonium for nuclear weapons. Since then, many countries have built such reactors, including two commercial ones in Germany and in the United States respectively, as well as test reactors in the United Kingdom, Japan, and China. Each of these reactors proved problematic, suffering a variety of failures and unplanned shutdowns.[10]

Leaks of various components resulted in oil or water getting into the nuclear cores of these reactors on multiple occasions. Water poses a special danger. When it enters the reactor core, the reactivity of the system goes up because the water slows down neutrons. That scenario constitutes one of the two chief accident sequences—the other being entry of air—involving high-temperature gas-cooled reactors that could lead to nearby populations being exposed to significant amounts of radiation.[11]

As a result of these failures and outages, all these high-temperature gas-cooled reactors operated for only a fraction of the time. A standard measure of performance called load factor for the four commercial high-temperature gas-cooled reactors range from a maximum of 62 percent for the AVR in Germany to an abysmal 15.2 percent for the Fort St. Vrain reactor in the United States. No wonder owners of these reactors shut them down well before their operating licenses expired.

The case of the Fort St. Vrain reactor reveals some of the challenges. Construction of the reactor started in September 1968 and the reactor reached criticality in January 1974, but its performance was so erratic that its owners waited for over five years before declaring it to be operating commercially. Between 1981 and 1989, the reactor suffered "279 unusual events" such as incursions of water and air, according to a 2003 report from the Oak Ridge Laboratory. There was also a major safety problem with its control rods whose function is to regulate the rate at which fission reactions occur in the reactor. In 1988, the plant's owners decided to shut it down, telling the *New York Times* that despite their efforts, "it seldom runs."

Even China, which has repeatedly shown that it can build infrastructure within budget and on schedule over the last quarter century or more, has struggled with high-temperature gas-cooled reactors.[12] The HTR-PM, China's demonstration high-temperature reactor, was to have been built between 2007 and 2010. Construction eventually started only in 2012. At that time, officials predicted starting operations within "50 months." Instead, the HTR-PM took ten years to reach full power, and, in the first few months after reaching full power, operated with a load factor of around 10 percent. That's even worse than Fort St. Vrain.

Sodium-cooled fast neutron reactors

Another old technology being sold as solving the problems of nuclear power is the sodium-cooled fast neutron reactor. Bill Gates and the climate scientist James Hansen, both of whom frequently extol nuclear energy as a solution to the climate crisis, ardently promote this type of nuclear reactor. Hansen has written about fast neutron reactors in glowing terms in his book *Storms of My Grandchildren*. For his part, Bill Gates has funded a company (TerraPower) and raved about these reactor designs often—for example, during a 2021 CBS interview.

But these hopes are misplaced. One obvious problem with these reactor designs is their use of sodium, which reacts violently with water and burns if exposed to air. Reactors cooled with sodium are thus susceptible to fires. Sodium can also interact chemically with the stainless steel used in various components of the reactor as the temperature of these components varies, a likely cause for practically all such reactors leaking sodium.[13]

Fast neutron reactors are also capable of a variety of accidents. There is a long history of such accidents starting with the 1955 partial core meltdown of the EBR-1 in Idaho, and the devastating accident in 1966 at the Fermi-1 demonstration fast reactor near Detroit, Michigan (see chapter 1). A particular concern with fast reactors are the so-called core disruptive accidents, where the core heats up, assumes a more critical configuration, and blows itself apart—a possibility first explored in 1956 by the Nobel Prize–winning physicist Hans Bethe. Having to deal with these safety concerns makes fast neutron reactors significantly more expensive to build than the more common thermal reactors, as the International Panel of Fissile Materials explained in a 2010 report on the topic.[14] (Full disclosure: I contributed a chapter on India's breeder reactors to the report.)

Nor are these reactors likely to be built quickly, and India's prototype fast breeder reactor (PFBR) offers an illustration of the lengthy delays associated with developing such designs. Plans for building the PFBR commenced in the early 1980s, after a quarter century of dreaming about breeder reactors. In 2004, when the first concrete was poured, the PFBR was expected to start operating in 2010. The reactor has been delayed repeatedly and the company building it claims it will start operating in 2024.

Countries around the world have pursued building such reactors, and the International Panel on Fissile Materials estimated the total spending at "tens of billions of dollars" in its 2010 report. Yet no breeder reactor has succeeded commercially. Nor, for the reasons mentioned above, should one expect good performance in the future.

Bill Gates might believe that his company will be different, but this is unlikely. How sodium behaves when it interacts with air or water does not change, even if the sodium resides within a nuclear reactor backed by one of America's oligarchs.

Molten salt reactors

Another class of problematic reactor designs rediscovered in recent years uses molten salts. This type has been pursued by Transatomic, before it folded up; two companies that have received Canadian government support, Terrestrial and Moltex; as well as Korean, European, and Chinese entities. As the name suggests, such reactors use nuclear materials dissolved in hot chemical salts. Thus, reactor components would operate within a chemically corrosive, hot and highly radioactive environment. Decades of search have identified no materials that can survive for long periods in such an environment without losing their integrity. A 2018 review from the US Idaho National Laboratory could recommend only that "a systematic development program be initiated."

The Molten Salt Reactor Experiment, which operated intermittently from 1965 to 1969 at the Oak Ridge Laboratory, offers the only real empirical experience with such reactors. Over those four years, the reactor was shut down 225 times; only fifty-eight of these shutdowns were planned.[15] The remaining were due to unanticipated technical problems. Even the US Atomic Energy Commission, which had funded the Oak Ridge reactor and related research for nearly two decades, raised difficult questions about the technology in a devastating 1972 report.

Those problems have not gone away. Numerous technological challenges remain to be overcome, concluded a 2015 report from France's Institut de radioprotection et de sûreté nucléaire. A few tens of millions of dollars in venture funding from a Silicon Valley investor will not make these hard technical problems disappear—certainly not within the short time frame we have to deal with climate change.

Safety, proliferation, and waste

All of these "advanced" or "generation-IV" or "small modular" reactors bring with them the usual predicaments. They are at risk of severe accidents, produce radioactive waste, and create the means to acquire nuclear weapons. As Edwin Lyman from the Union of Concerned Scientists explained in great detail in his comprehensive 2021 report, "advanced" isn't always better.

Reactor designers can set out to address these problems. But it is impossible to address all these goals concurrently, especially when also trying to produce a design that can compete economically with alternative ways of producing electricity.[16]

Consider, for example, how to lower the amount of radioactive waste produced. Technically, this would require the nuclear reactor to utilize fast (i.e., more energetic) neutrons. Any such reactor would produce more plutonium for each unit of electricity generated; more importantly, spent fuel will contain a higher concentration of plutonium. As a result, would-be proliferators will need to divert less spent fuel to a (possibly clandestine) reprocessing plant to produce plutonium to make nuclear weapons. Thus, such a reactor design increases the risk of proliferation.

Risk of accidents

As I described in chapter 1, nuclear reactors are inherently hazardous, because of fundamental technical reasons. All nuclear plants, including any involving these new designs, involve immense complexity and rapid processes. The possibility of accidents resulting in widespread radioactive contamination is part and parcel of this territory.

Different reactor designs may be susceptible to, or immune to, different sequences of events that lead to such accidents. A sodium-cooled fast reactor, as TerraPower and Bill Gates keep emphasizing, is not at any risk due to steam pressure building up, because it is cooled by sodium. But such reactors can undergo core disassembly accidents that could result in explosions. The explosion would overwhelm the containment building, allowing for radioactivity to escape into the atmosphere.

All else being equal, a smaller reactor is better because of its smaller inventory of radioactive material; it will also have less energy available for release during an accident. But smallness does not offer a panacea. Even in a very small microreactor (say, one that generates under ten megawatts of electricity), a severe accident can release enough radioactive materials to expose members of the public to significant radiation doses, according to calculations in an April 2020 report from the Idaho National Laboratory.

At the same time, the smaller power output will necessitate building more of these reactors to make even a dent in global emissions. More reactors means more locations for potential accidents.

In the real world, however, all else is seldom equal. Power plant designers also have financial priorities. This is why small modular reactor proposals often envision building multiple reactors at a site. Taking advantage of common infrastructure elements, the hope goes, could lower unit costs. NuScale, for example, plans to build twelve reactor modules at each site. With multiple reactors, the combined radioactive inventory becomes comparable to that of a large reactor.

Multiple reactors at a site also increase the risk of contagion: an accident at one unit might induce accidents at other units. This is even more probable when the underlying reason for the accident is a common one that affects all the reactors, such as an earthquake. In the case of the accidents at Japan's Fukushima Daiichi plant, explosions at one reactor damaged the spent fuel pool in a co-located reactor.

Accidents at one unit also make it harder to take preventive actions at other units. Radiation leaks from some of the

Fukushima Daiichi reactors made it difficult for emergency workers to approach the other units.

Radioactive waste

All nuclear reactors, simply by virtue of the underlying fission process, will necessarily produce radioactive waste. There is no demonstrated solution to dealing with this by-product (see chapter 1).

Nuclear advocates have a split attitude toward radioactive waste. On the one hand, they trivialize the problem. The Nuclear Energy Institute, for example, says, "The entire amount of waste created in the United States would fill one football field, 10 yards deep." In other words, no big deal.

At the same time, many companies promote nuclear reactor designs fueled with plutonium extracted from waste. For example, a company called Moltex has received over $50 million in funding from Canada's federal government. The government justified the funding to help "reduce storage of nuclear waste." And Oklo, a company going public using the SPAC route in the United States, advertises its reactors as running "off nuclear waste."

The claim about operating on waste is disingenuous, since the unsaid implication is that no radioactive materials will remain. The nuclear fission process underlying energy production in any reactor will always results in a great many radioactive substances—fission products. None of these substances can undergo fission to produce more energy. Operating reactors also contain radioactive activation products when structural materials—steel, for example—absorb neutrons. These, too, cannot undergo fission. Only a few radioactive substances—plutonium, for example, produced when uranium nuclei absorb neutrons—can undergo fission and thus fuel another reactor.

The remaining radioactive materials must be managed in one

way or the other. Some of these materials have a short half-life and they rapidly decay into stable elements. But many elements will be radioactive for much longer periods.

In the case of the Moltex reactor design, my friend the nuclear engineer Jungmin Kang used a standard computer code called Origen-2 to calculate what would be left behind as waste, while producing the initial batch of fuel loaded into a single 300-megawatt Moltex reactor. According to these calculations, producing the initial core for a reactor using Moltex's proposed process would result in radioactive waste forms containing 113 terabecquerels (or trillion becquerels; one becquerel corresponds to one nucleus disintegrating every second) of technetium-99 (half-life of 211,000 years); 225 gigabecquerels (or billion becquerels) of iodine-129 (half-life of 15.7 million years); 825 gigabecquerels of cesium-135 (half-life of 2.3 million years); and 14.9 terabecquerels of carbon-14 (half-life of 5,700 years)—these radioactive levels are calculated for a thousand years into the future, when short-lived radionuclides would have decayed away.[17] Over its lifetime, producing fuel would result in ten times these quantities. These large inventories of radioactive materials mean that wastes from Moltex will remain hazardous for millennia.

Because of their smaller size, small modular reactors would produce more wastes of different kinds relative to currently operating gigawatt-scale reactors when these two reactor designs are weighted by how much electrical energy they produce. The increased production of radioactive materials is an example of the how a gain in one metric (safety) is accompanied by a loss in another metric (waste production). Likewise, "advanced" reactor designs that are not cooled with water will necessarily involve materials that are more problematic to manage, as my colleague Allison Macfarlane and her former postdoc Lindsay Krall have argued. Some of these materials are corrosive while others are pyrophoric. Dealing with them is complicated. For example, fast neutron reactors, such as the one proposed by Bill Gates, will use sodium, and sodium, as described earlier, reacts strongly with water. So, the waste generated in such reactors cannot be disposed of in geological repositories without extensive processing. Such processing has never been carried out at scale.

Nuclear weapons proliferation

As discussed in chapter 5, expanding nuclear energy automatically increases the risk of more countries acquiring nuclear weapons. Should the global market for nuclear power involving these new reactor designs expand rapidly, as proponents claim, then many countries that are currently without requisite capabilities will acquire some of the technical means to make nuclear weapons.

Over and above the general linkage between nuclear weapons and nuclear energy technology, some of the specific designs pose particular risks. Most of these reactors use fuel forms involving either enriched uranium or plutonium, often at concentrations greater than in currently operating reactors. Nuclear weapons can use both plutonium and uranium highly enriched in the uranium-235 isotope.

All else being equal, nuclear reactor designs requiring fuel with higher levels of uranium-235 increase the risk of nuclear proliferation.

All nuclear power plants using uranium fuel produce plutonium, but only alongside intensely radioactive fission products. To fabricate nuclear fuel, the plutonium must be partitioned from these materials—such a process will automatically make it easier to use this material to make nuclear weapons. Practically any mixture of plutonium isotopes is usable in nuclear weapons.

I should also say a word about thorium here: no. Thorium reactors are not free of risk. As discussed in chapter 5, in order to operate reactors, the thorium has to be converted into uranium-233, and nuclear weapons can be made with uranium-233 just as well as they can utilize plutonium.

The bottom line: the old problems of nuclear energy will also afflict new reactor designs. There is no escape from the laws of physics, chemistry, and biology.

Sales tactics

Given these problems, how do investors sell new nuclear reactor technologies?

Reading news stories related to this topic, one strategy becomes quickly apparent. Reactor design vendors and those investing in these companies put out a lot of "happy talk" (the term for optimistic and exaggerated claims that I first heard from Vinod Mubayi, a physicist who consulted for the Nuclear Regulatory Commission).

Promoters claim just about everything one might want from nuclear reactors. Their products are simple, safe, cheap, and quick to build; they will reduce how much radioactive waste is produced or even reuse waste; they can be exported to lots of countries without giving them any way to make nuclear weapons; and yes, they solve climate change. As Benjamin Sovacool and I have argued at some length, these claims are packaged into several utopian fantasies and visions, reminiscent of the nuclear dreams that proliferated in the immediate aftermath of the bombings of Hiroshima and Nagasaki.[18]

The most important of the rhetorical visions that Benjamin and I identified in our analysis of claims about small modular reactors were those involving "risk-free energy" and "environmental nirvana." The first vision involved the claim that new nuclear reactors would produce "energy with perfect reliability and complete safety," and the second vision posited that these machines would "deliver clean and plentiful electricity in a carbon constrained future."

These are extraordinary claims and, as such, need extraordinary evidence to render them credible. As I have explained at length, there are historical and technical reasons why these claims are not true. Yet because nuclear advocates constantly repeat these talking points, they have become ubiquitous and have been adopted by policymakers.

Media houses have acted as a tool to promote new reactor designers. Increasingly, articles use "is" or "are" where "could be" might offer the most charitable description. For example, a 2019 article published by the Canadian Broadcasting Corporation about the nuclear industry's pitch that SMRs could make "Alberta's oilsands cleaner" announces: "The reactors *are* cheaper, can be built out of standardized units created in factory and assembled on site" (my emphasis). Well before 2019, plenty of studies showed that SMRs would not be cheaper but more expensive.[19]

In the case of Transatomic, news articles uncritically reproduced the developers' claims about safety and waste reduction. During Bill Gates's frequent appearances on TV channels, hosts never questioned him about the sorry history of fast neutron reactors. The media has been complicit in promoting unwarranted excitement about small modular and advanced reactors.

Another tool in the nuclear PR effort comes from Silicon Valley: virtual simulations. In a 2001 article in *Insight*, Jill Clelland, a lead manager at Westinghouse Electric, described how the company utilized computer-based tools—developed using public funding at a public university—to execute a "virtual construction project." The "project team" Clelland writes, "not only *verified* the schedule through visualization, but it found another three to five months could be eliminated through logic and design changes alone" (my emphasis). Note the inappropriate use of the word "verified" for what came out of a computer simulation.

When Westinghouse's AP1000 reactor was finally built, the actual construction time exceeded the simulated projected schedule by over a factor of three. The failure in the real world did not result in the nuclear industry discontinuing reliance on such tools. If anything, it has doubled down on such simulations. The reason: simulations can help persuade potential investors and clients by creating (false) confidence. As Clelland explained, "Visualization offers tremendous sales potential."

One institutional player has not quite fallen for the sales pitch, at least not just yet: the safety regulator. Regulators, too, feel the attraction of overseeing a large expansion of nuclear power. The Canadian Nuclear Safety Commission, for example, has promoted Canada as a preferred destination for small modular reactors by highlighting how its approach to licensing reactors differs from the US Nuclear Regulatory Commission's methods.[20]

Yet their mandate requires regulators to critically question the safety claims made by companies. Such questioning could pose problems for some companies. In January 2022, the US Nuclear Regulatory Commission rejected an application from Oklo to license its Aurora reactor. The NRC explained that Oklo had "repeatedly failed to provide substantive information in response to NRC staff requests for additional information ... on the maximum credible accident ... for the Aurora design" and "the safety classification of structures, systems, and components." Although impossible to know for sure from the outside, the plausible reason for Oklo not supplying the required information is that the company has simply not performed the needed safety analyses.

The decision drew loud complaints from nuclear proponents. The Nuclear Innovation Alliance, a think tank, told CNBC that "the decision was a disappointment and a sign of outdated regulatory processes" adding the wishful claim that advanced "reactors are expected to be safer than any reactors to date and should be able to meet NRC's standards."

Complaints about the NRC are not new. Back in 2016, Jacob DeWitte, Oklo's CEO, told the US Senate's Committee on Energy and Natural Resources: "The regulatory process as it exists today is not well suited for these new technologies and the venture finance models that fund them."

In their ideal world, such developers want regulators to simply abdicate their jobs. Speaking to *CoinDesk*, Ed Lyman from the Union of Concerned Scientists explained that Oklo, and some other new reactor companies, "just want the NRC to accept the reactor is going to be safer" and "essentially let them do whatever they want." As the Transatomic case shows, the developer's assurances about the reactor's performance could be dubious; identifying such mistakes is one purpose of safety reviews.

Letting companies "do whatever they want" is, however, the norm in some circles. In an interview with Y Combinator, the venture capital organization, DeWitte explained that "the receptiveness to what we were doing was so different out in Silicon Valley than it was on the East Coast. On the East Coast, we'd often be met with skepticism, people asking, 'Is that safe? How is that possible?' Out here, it was like, 'How can I help?'"

The political right also loves the idea of scrapping regulations. As an example, consider the Heritage Foundation's "policy briefing guide" called Solutions, which "offers conservative recommendations on key policy issues." The guide's proposals for energy policy include: "Stop the regulation of greenhouse gases" and "Overhaul nuclear energy regulation."

Finally, there are nuclear boosters like Ted Nordhaus (see chapter 5), a signatory of the Ecomodernist Manifesto, who denounced the Nuclear Regulatory Commission for "the decline of the legacy nuclear industry" in an April 2023 article with Adam Stein in *Foreign Policy*. In other words, if nuclear energy has problems, it must be the fault of regulators.

Criticizing regulations has been part of mainstream political culture since Reagan and Thatcher. So, bipartisan support for whittling the NRC's power is expected. As mentioned in chapter 1, during the 1990s Senator Pete Domenici forced the agency to ease regulations by threatening to cut its budget. More recently, the 2019 Nuclear Energy Innovation and Modernization Act signed by former president Donald Trump forces the NRC to "reform" its fee structure and "develop a streamlined licensing process for advanced reactor designs." "Reform" and "streamline" are code words—the nuclear industry and its friends are forcing the NRC to reduce its questioning and charge companies less, thus weakening its capacity to regulate.

Nuclear companies developing new reactor designs also adopt standard organizational techniques to help with accessing political power and capital. Early in this chapter, I described the parade of notables promoting or investing in Transatomic prior to its shutdown. Such involvement is common. In Canada, two companies are peddling molten salt reactor designs: Terrestrial Energy and Moltex. Terrestrial's list of advisors features people like Stephen Harper, former prime minister of Canada; Ernest Moniz, former US secretary of energy; Ray Johnson, former chief technological officer for Lockheed Martin; Lord John Browne, former CEO of BP; Ray Rothrock, the venture capitalist mentioned earlier; and Robert Litterman, former head of risk at Goldman Sachs. Likewise, Moltex's Advisory Committee includes representatives of the three corporations that operate nuclear power plants in Canada: Brett Plummer, vice president of nuclear and chief nuclear officer of NB Power; Dominique Minière, president of nuclear of Ontario Power Generation; and Mike Rencheck, president and CEO of Bruce Power.

Shaping the future

Every week, my Google Alert for the term "small modular reactors" feeds me dozens of articles on the subject. Without the benefit of my years of research into nuclear power, reading these articles regularly would have convinced me that a new era of nuclear power is imminent, and will soon solve climate change, illuminate houses of poor villagers, produce hydrogen cheaply, and provide water by removing salt from seawater. I expect that one of these days, one of these will promise to slice bread too. In this avalanche of propaganda, honest assessments of the prospects of these reactors and their ability to deliver the promised benefits are as scarce as hen's teeth.

I get it. Journalists writing these articles face tight deadlines. So, I imagine they rely primarily on the public relations materials circulated by companies selling nuclear energy. What I find less obvious is why investors are putting millions of dollars or euros into patently problematic technologies.

My arguments about why fancy qualifiers like "advanced" or "innovative" don't prevent the many problems with older nuclear technologies from afflicting these reactor designs are not new. At the risk of immodesty, I can claim to have written extensively on this subject. This book's notes at the end include references to many of my articles.

Nor am I the only one doing so, not by a long shot. Many of the excellent scholars, analysts, and professionals mentioned in this book have produced comprehensive and wide-ranging critiques.

This literature is easily available to anyone who wants to dig deeper. One would think that Silicon Valley's billionaires would engage in due diligence before investing their money. Instead, their actions are better described by what Walter Bagehot, editor of the *Economist*, wrote about the South Sea Bubble of the early eighteenth century: "At particular times a great many stupid people have a great deal of stupid money."

With unrivalled fortunes, many individuals investing in nuclear reactors have ample wealth and can afford to lose millions on various unprofitable ventures. Society's problem is what else they do: using their public reach to hype nuclear energy, especially the specific nuclear reactor they are investing in, as well as getting governments to channel public money into nuclear companies.

In recent decades, "the cult" of the entrepreneur has risen to "near-mythic prominence," as Shoshana Zuboff argued in *The Age of Surveillance Capitalism*. So it shouldn't surprise one that both the public and policymakers are receptive to these pronouncements, even though they are frequently disproved. Elon Musk's many predictions about Tesla's self-driving cars and robotaxis are an example.

The greater peril resulting from this phenomenon is the propensity of this "new generation of storytellers," as Nicole Aschoff aptly describes them in her 2015 book *The New Prophets of Capital*, "to tell us what's wrong with society and how to fix it." These stories, Aschoff argues, "set the terms of debate and the field of possibility, dominating the plane of ideas and swallowing up stories that challenge the status quo. Their stories enable capitalism to evolve and absorb critique, thus preserving itself as a system."

But stories also do something much more material for these billionaires. As venture capitalist Ray Rothrock told the US Department of Energy's Office of Nuclear Energy, even in the case of a risky project, "a great story" increases the confidence of the investor. Rothrock also explains his theory of how to deal with the risk: "Two things motivate people; fear & greed, but greed is a much more powerful motivator. And to talk about it that way is an opportunity." As we have seen in this chapter and elsewhere in this book, there are many ways in which nuclear energy feeds the greed of investors, including those whose money went into the $9 billion–plus failed nuclear project in South Carolina (see chapter 3).

For some of these investors, the fear that their investment proving unprofitable might be counteracted by another fear. Speaking to CNBC about going public using the SPAC route, Sam Altman, CEO of OpenAI, explained how "the future can be radically better" provided one could "lower the cost of energy and lower the cost of intelligence." These two are related in Altman's mind because the computers used for artificial intelligence predictions will need a lot of energy.

But after connecting his investment in nuclear power with his artificial intelligence interests, Altman went on to his deeper concern: "'The alternative to not having enough energy is that crazy de-growth stuff people talk about. We really don't want that,' referring to the philosophy that restricting production, consumption and energy use is a way to conserve natural resources. 'I think it's insane and pretty immoral when people start calling for that.'"

Altman's resort to name calling suggests that he might be worried by the idea of degrowth catching on. Those concerned with environmental problems and climate change might start seeing the absolute necessity for systemic change rather than superficial technological bandages. Any such change would affect the privileges of Sam Altman and the small group of insanely wealthy people who occupy the 0.1 percent of today's wealth bracket.

Conversely, promoting nuclear power and other untested technologies serves to divert the public's attention away from the larger systemic drivers— in particular, unabated capitalism and its need for never-ending economic growth—of the climate crisis. Pushing the nuclear agenda allows maintaining the false idea that the current pattern of development can continue indefinitely with no limits, while climate change is solved by using one more technology from the same toolbox responsible for the problem in the first place.

Conclusion

When they don't know what to say
and have completely given up on the play
just like a finger they lift the machine
and the spectators are satisfied.

Antiphanes

Technology discloses man's mode of dealing with Nature, the
process of production by which he sustains his life, and thereby
also lays bare the mode of formation of his social relations, and
of the mental conceptions that flow from them.

Karl Marx, *Capital*[1]

Nuclear energy cannot save us from the climate crisis, and building more reactors would give rise to a number of undesirable consequences. These consequences include the increased risk of radioactive fallout from severe accidents contaminating large swaths of land; growing inventories of radioactive wastes of different kinds that are difficult to manage because they remain hazardous for hundreds of thousands of years; the likely spread of the capacity to produce materials used in nuclear weapons; and in countries that already have such weapons, a further consolidation of the economic and political power of the institutions involved in making them.

Severe accidents are made more likely because nuclear facilities are operated by organizations with multiple priorities, including cost-cutting and profit-making; we cannot rely on regulatory oversight to ensure safety. Because of the special challenges associated with radioactive waste, the nuclear industry's preferred solutions to managing its wastes, such as burying these in geological repositories or reprocessing them to extract plutonium, are incapable of assuring that radioactive materials will not contaminate the biosphere in the short or long run. The nuclear energy and weapons enterprises are intimately related due to overlaps in the necessary technologies, personnel and training, and institutional interests. The ability to use nuclear energy for military purposes is an asset for nuclear advocates, especially when they seek government support, and the continued use of nuclear reactors to generate electricity is a major obstacle to global nuclear disarmament, the only way to eliminate the risk of catastrophic nuclear war.

Government support is critical to nuclear power. Governments provide subsidies and justify such funding by making groundless assertions about nuclear power's environmental desirability or its economic attractiveness. Governments also create the legislative environment necessary for private corporations building or operating nuclear power plants to socialize their costs and risks and to privatize their profits. Because profits are a priority, and there is often a confluence of interests between owners of nuclear plants and fossil fuel plants, who are often the same entities, organizations invested in nuclear energy are unlikely to support rapid and wide-ranging action on the climate.

Finally, supposedly new kinds of nuclear reactors are not immune to the risk of severe accidents, will continue to produce radioactive wastes, and will continue to be connected to the possibility of producing materials to use in nuclear weapons. Despite these problems, wealthy businesspeople like Bill Gates promote these designs as innovative. Although they invest in companies developing such *faux* nuclear plants (see the introduction), they invest only small amounts relative to their individual wealth and then go on to seek public money. They also blame regulators for being too strict and stifling innovation, but lax regulation can make accidents only more likely.

When coupled with Silicon Valley's slogan "Move fast and break things," such investors are certain to leave the public with a radioactive mess.

For those who might feel that such problems with nuclear power are to be tolerated because it offers a solution to climate change, I have shown why it is infeasible for nuclear energy to expand quickly enough to meaningfully help lower carbon dioxide emissions to levels sufficient for keeping global temperatures under 1.5 degrees Celsius. It would take too long to build the thousands of nuclear reactors that proponents of the technology are calling for to deal with the climate crisis. It takes fifteen to twenty years to plan for and build each nuclear plant, and even longer for unproven theoretical reactor designs, such as small modular reactors or high-temperature gas-cooled reactors. Over this period of time, they produce no electricity, and won't contribute to reducing emissions, but will divert financial investments and people from more sustainable energy technologies. In contrast, renewable sources of energy like solar and wind power can be scaled up rapidly and thus will reduce emissions much more quickly.

This prognosis is for countries with ample experience at building nuclear plants. For example, in the United States, the country with the most operating reactors, the most recent reactors built have taken at least eighteen years to go from the planning stage to supplying power to the grid. Planning for Vogtle-3 and Vogtle-4 reactors started in 2005, during the heyday of what was promised to be a nuclear renaissance, and the first of these units began operating in 2023. France, which is more reliant on nuclear energy than any other country, has taken even longer to plan and build its latest reactor, Flamanville-3.

For nuclear power to contribute meaningfully to climate mitigation, however, a large fraction of new nuclear plants would need to be built in developing countries; it is in such places that the demand for electricity has been growing most rapidly. In these countries, the timeline for building a nuclear plant will be much longer than in the United States or France, which have, after all, built dozens of nuclear plants over the decades. The timeline would be further challenged in countries confronting greater political instability or possessing weaker state capacity.

In all countries, moreover, nuclear plants are among the most expensive ways of producing electricity for the grid. Spending money on building them will necessarily take money away from other sources of energy, especially renewables. Such spending will particularly affect cash-strapped developing countries and impede their ability to provide electricity for people that don't currently have access. Renewables would meet their needs more cheaply and quickly.

This is an open and shut case. Expanding nuclear energy will not help address climate change but will worsen a range of environmental problems and security risks.

Dealing with renewables

Advocates of nuclear power object to being compared with solar and wind energy. They might ask: How can you maintain an electrical grid with renewables when there are times that the sun is not shining and the wind is not blowing? Some version of this question has come up after many of my seminars and lectures.

In my experience, this question typically serves only a rhetorical purpose. The questioners are usually not interested in understanding how a grid works, and how it could be modified to accommodate the inevitable increase of variable renewable sources of energy. They are just seeking to make the case for nuclear power or even natural gas plants.

Those who are asking it sincerely, however, are mostly drawing on mainstream sources of information that regularly emphasize the need for something called "baseload sources of power." That term derives from an era when most of the demand for electricity was met by fossil fuel sources or nuclear energy. In that way of thinking, which has a hegemonic status among energy planners, the backbone of the electricity grid is constituted by power plants that generate electricity at a more or less constant level of output: twenty-four hours a day, 365 days a year. These so-called baseload sources traditionally included coal power and nuclear power plants. In more recent decades, combined cycle natural gas plants have joined this category of electricity generators.

Baseload plants provide a constant output. But they have to be complemented by other power plants that supply electricity during those hours of the day or times of the year when demand is higher than what baseload plants can supply. Examples of higher-demand periods include hot summer days when lots of air conditioners are operating, or early evening hours when people come home and switch on their televisions and microwave ovens. To meet the added need for electricity during those specific, and often short, periods of time, utilities have usually relied on power plants that operate just during those periods. Such plants must increase and decrease their outputs rapidly. In energy-planning jargon, these are "load-following plants." Traditionally, the function of load following is provided by natural-gas- or oil-based "peaking plants."

Nuclear plants are better suited to baseload generation than to load following, as using them in the latter mode has technical and economic implications. Abrupt shifts in the levels of power generated can result in the temperature of the fuel changing quickly and, in turn, increase the likelihood of radioactive materials leaking out. Safety concerns, thus, limit the ability of nuclear reactors to complement solar and wind power, the outputs of which can change quickly.

It is possible to design nuclear power plants differently, but there is still an economic problem. Nuclear power plants have high capital costs associated with construction, but low fueling costs. Therefore, there is a financial incentive to operate them continuously, at near their maximum capacity. This is why nuclear power plants have been traditionally treated as a baseload electricity source. Oil- or gas-fired plants are a better choice for covering peak electricity demand, because of their low capital and high fuel costs.

Modern renewables change this picture in an important way. Their outputs are determined by natural forces that are beyond the power of corporations or individuals to control. They can be classified neither as baseload nor as peaking plants. As more and more solar panels and wind turbines are put up, their combined outputs can sometimes dominate electricity supply especially when demand is low. There may even be periods of time when the entire demand is met by solar or wind power or some combination thereof.

One indication of the changing nature of the electricity market is the increasingly common occurrence of negative price bids on the energy spot market. This counterintuitive phenomenon occurs when owners of baseload power plants would rather pay other power plant owners to stop generating electricity rather than shutting down their baseload plants, because of the economic cost of shutting down and restarting their plants and because such ramping up and down of their baseload plants increases the wear and tear on some of their components.

The growing capacity for renewable energy, therefore, requires a shift in the basic paradigm that has governed electricity planning for close to a century, which classifies generation capacity into baseload and peak. To integrate more renewable energy technologies, the grid would need more flexible sources of power that can quickly vary their outputs in response to changes in the outputs of renewables. While fossil fuel–based peaking plants could do that job, their use would not fit the goal of reducing greenhouse gas emissions.

Planning for such a grid will also require new ways of thinking, a new paradigm. Transitioning to a new paradigm will not be easy, even for those who do not have serious economic interests in maintaining nuclear or coal energy. Thomas Kuhn discussed the key challenge in such a transition in *The Structure of Scientific Revolutions* (1962): those trained in one paradigm—energy planners in this case—find it very difficult to think outside that paradigm. It is this conceptual inertia that is on display when energy analysts repeat the mantra that we need baseload sources of power.

A new paradigm is yet to evolve fully, but there are many people who have spent years thinking about how to manage a grid with variable sources of renewable energy. Here, I summarize briefly what I understand of these efforts.

One can start by remembering that even during earlier decades, when there was little or no solar and wind power being produced, grid managers had to contend with variations in the supply of electricity. Nuclear plants, for example, have to be shut down for refueling. Water availability varies with the seasons, meaning the power produced by hydroelectric dams also changes over the year. These are predictable variations.

Grid managers have always had to deal with unpredictable disruptions too. A coal plant's fuel supply might be interrupted because a train derails, for example. A nuclear plant, or even multiple nuclear plants, might unexpectedly need to be shut down for safety or other reasons. (French nuclear plants were, on average, shut down for 96.2 days, 115.4 days, 103.8 days, and 152 days in 2019, 2020, 2021, and 2022.) Unanticipated shutdowns will become more common as the frequency of climate change–linked extreme weather events increases.

Thus, managing grids will inevitably involve managing unpredictable variation in supply, regardless of renewables being added. The variability in power from solar and wind, of course, is different from the kind of outages seen in nuclear and coal plants. It depends more sensitively on weather patterns.

With improved weather forecasting, though, it has become easier to predict how much power will be available in the future—say, twenty-four hours ahead. There are, broadly speaking, three ways to address this variability. The first approach is to increase diversity in technologies and in geographical locations. There are many technologies lumped under the term "renewables": onshore wind, offshore wind, solar photovoltaics, solar thermal, geothermal, and small-scale hydropower. (Because of their sizable impacts, large hydropower dams are usually not included under modern renewables.) Not all of them have seen the dramatic declines in cost witnessed in the case of solar photovoltaics and wind turbines, but they could all be deployed at some level, depending on the location. For example, in his 2016 study of a renewable energy system for the state of Maryland, energy analyst Arjun Makhijani highlighted the seasonal balance between solar energy (plentiful in summer, but not in winter) and offshore wind energy (more plentiful in winter). Geographical diversity also helps. If the wind is not blowing in Germany, it could be blowing in Spain.

The second approach is to change the patterns of electricity demand. This could be done by consumers shifting their electricity use to periods when renewable energy is plentiful. Utility companies have used price signals as incentives for this purpose, albeit with unequal impacts on poorer consumers. But consumption shifts could also be concentrated on commercial and industrial sectors. Demand shifts could be combined with measures to encourage much more efficient use of energy. A building that gains or loses heat more slowly can remain comfortable for longer periods of time, even during periods of reduced electricity supply. There is also a larger conversation to be had about whether some forms of energy use should be totally prohibited; powering computers to do pointless computations in support of cryptocurrencies, for instance.

A final option for dealing with variability is through energy storage. There are multiple forms of energy storage, from lithium-ion batteries to pumped storage of water. Many are rapidly declining in cost, and they are being deployed rapidly. In October 2023, for example, the state of California announced that it had installed more than 6,600 megawatts of battery energy storage systems, up from 770 megawatts in 2019.

How much these storage technologies can economically contribute to enhancing the reliability of the grid is a question for the future. Historically, the energy policy community has been terrible at predicting future costs. Solar and wind technology costs have become lower than experts had predicted; nuclear costs have been significantly higher. There is good reason to be skeptical of long-term predictions of the cost of batteries and other forms of storage.

As a result of such uncertainties, predictions for the most economical way to supply electricity reliably in the far future will always be dubious. The far future is the appropriate time frame to pose this question because the share of renewables in most countries is unlikely to reach high enough levels to pose significant challenges to reliability anytime soon, at least if one goes by past trends. For the present or the near-term future, there is little doubt that the focus of expanding energy supply should be on renewables and associated technologies.

There is a legitimate debate to be had about whether nuclear energy is needed to balance the variability of renewables, but that debate also has a legitimate conclusion: there is no evidence for nuclear power being needed to make the grid reliable.

The relationship of energy to political economy

Renewables are no panacea. Just like nuclear reactors and fossil fuel plants, they do have their own environmental and public health consequences, resulting from the material requirements for solar panels, wind turbines, and batteries. The demand for these materials leads, in part, to the exploitation of miners in the Congo and the contamination of water in South America. Once again, there is no need for me to expand on this—plenty of articles in mainstream media routinely expose one or more of these tragedies.

These injustices serve as a reminder that a truly sustainable and just energy system should incorporate multiple goals, addressing injustices such as local pollution (including from the manufacture of renewables) and inequality in access to energy. Green capitalist strategies that focus only on carbon dioxide emissions while relying on market forces to drive renewable technologies are likely to exacerbate tensions between these multiple goals.[2]

How renewables impact people in countries like the Congo is completely predictable in light of what I have described earlier about how the communities most often subject to the environmental and public health impacts of the nuclear industry are politically weak, economically poor, and geographically far from the corridors of power. These vast disparities of wealth and power in contemporary society are, arguably, as important to the enterprise of generating nuclear power as the capacity for high-tech manufacturing or uranium ore.

How the nuclear industry functions, including its relationship to the capitalist system and what it reveals about the drivers of the climate crisis, also points to the limitations of relying on renewable technologies alone to come up with ecologically sustainable and socially just energy systems. As the quote from Karl Marx's *Capital* in the epigraph of this conclusion suggests, the attraction of nuclear technology, despite its well-known problems, to elite decision makers discloses much about contemporary society. These features are deep-seated and unlikely to be remedied by a mere cataloging of the problems of nuclear power or positing renewables as the alternative. There are no solutions that will allow everything else to stay the same, from the energy we use to the financial structures that govern its production.

Absent fundamental changes, the impacts of the extractive processes associated with any form of electricity generation will only continue to accelerate. Capitalism is based on continuous economic growth. For this continuous economic growth to take place, capitalism needs energy and materials to feed ever-expanding production. But material and energy use cannot keep growing continuously on a finite planet. Renewables might not produce carbon dioxide when they generate power, but they do have their own material requirements.

The counterargument from mainstream policymakers, economists, and even quite a few environmentalists is that the economy can grow without a corresponding increase in emissions and energy use. This wishful hope is termed "decoupling" in the academic literature. Because of the concern about climate change, most studies of this topic have focused entirely on carbon dioxide and not on total material use. But even within this narrower field of inquiry, the evidence does not offer much hope. At the risk of simplifying a vast literature, decoupling of carbon dioxide emissions from economic growth does not seem to be happening at a global scale, certainly not at a rate that is relevant to climate mitigation.

This is an inconvenient truth. Although climate change has been widely recognized as a major problem, few confront that problem by examining and critiquing the globalized capitalist system that is driving it, in an unending search for profit through extraction, production, and ever-increasing consumption. Fewer still call for changing that economic system. The problem was well characterized by the cultural critic Fredric Jameson when he wrote on the pages of *New Left Review* in 2003: "Someone once said that it is easier to imagine the end of the world than to imagine the end of capitalism." That lack of imagination is a critical element in the story of nuclear energy's survival.

Nuclear energy is being promoted by powerful elites in governments and businesses precisely because it comes with the promise, even if it will be ultimately a false promise, that the economic system can continue more or less along the same path while avoiding large-scale climate change. It is analogous to the magician's misdirection, getting people to focus on shiny new technological solutions promised for the (distant) future, while business as usual goes on. Talking about nuclear power from new reactors serves to delay dealing with the climate crisis. Procrastination might be the thief of time, but it is good business strategy for companies that profit from the current system.

Therefore, over and above its many undesirable impacts, one further reason to resist a nuclear expansion is to limit the power of these elites and their wealth. An energy transition can serve as a lever to constrain the power of capital, which is essential if we are to transcend the present expansionary system and avoid being overwhelmed by the multiple ecological crises confronting us today. However difficult this challenge may appear to us, we will do well to remember James Baldwin's exhortation: "Not everything that is faced can be changed; but nothing can be changed until it is faced."[3]

In the face of the climate crisis, determining how energy is produced and used is one of the great struggles of our time. It is not the only struggle, but its conclusion will nonetheless affect the terrain on which many other struggles will be fought. If this book helps clarify the nature of that energy system, through the examination of one element—nuclear power—then it will have served its purpose.

Notes

Introduction

1 My calculations, based on data from Energy Institute, "Statistical Review of World Energy 2023," June 2023, energyinst.org.

2 Mycle Schneider and Antony Froggatt, *The World Nuclear Industry Status Report 2022* (Paris: Mycle Schneider Consulting, October 2022), worldnuclearreport.org.

3 Alexander Glaser, Laura Berzak Hopkins, and M. V. Ramana, "Resource Requirements and Proliferation Risks Associated with Small Modular Reactors," *Nuclear Technology* 184 (2013): 121–9, tandfonline.com.

4 M. V. Ramana and Zia Mian, "One Size Doesn't Fit All: Social Priorities and Technical Conflicts for Small Modular Reactors," *Energy Research and Social Science* 2 (June 2014): 115–24; this list of four problems is from Stephen Ansolabehere et al., *The Future of Nuclear Power* (Cambridge, MA: Massachusetts Institute of Technology, 2003), energy.mit.edu.

5 M. V. Ramana and Ali Ahmad, "Wishful Thinking and Real Problems: Small Modular Reactors, Planning Constraints, and Nuclear Power in Jordan," *Energy Policy* 93 (June 2016): 236–45, science direct.com; M. V. Ramana and Priscilla Agyapong, "Thinking Big? Ghana, Small Reactors, and Nuclear Power," *Energy Research and Social Science* 21 (November 2016): 101–13; Bernadette K. Cogswell et al., *Nuclear Power and Small Modular Reactors in Indonesia: Potential and Challenges* (Jakarta, Indonesia, and Berkeley, CA: Indonesian Institute for Energy Economics and Nautilus Institute for Security and Sustainability, April 2017), nautilus.org.

6 Some examples are Arjun Makhijani, *Light Water Designs of Small Modular Reactors: Facts and Analysis* (Takoma Park, MD: Institute for Energy and Environmental Research, 2013), ieer.org; Edwin Lyman, *Small Isn't Always Beautiful: Safety, Security, and Cost Concerns about Small Modular Reactors* (Cambridge, MA: Union of Concerned Scientists, September 2013); Mark Cooper, "Small Modular Reactors and the Future of Nuclear Power in the United States," *Energy Research and Social Science* 3 (September 2014): 161–77; Edwin Lyman, *"Advanced" Isn't Always Better: Assessing*

the Safety, Security, and Environmental Impacts of Non-Light-Water Nuclear Reactors (Cambridge, MA: Union of Concerned Scientists, March 2021), ucsusa.org.

7 Ross Carper and Sonja D. Schmid, "The Little Reactor That Could?," *Issues in Science and Technology* 84 (2011), issues.org.

8 "Industrial Africa with Atom Aid Seen," *New York Times*, January 28, 1947, 16.

9 These have been understood for decades. See, for example, Walter Rodney, *How Europe Underdeveloped Africa* (New York: Verso, 2018 [1972]).

10 David Beers, *Blue Sky Dream: A Memoir of America's Fall from Grace* (New York: Doubleday, 1996), 22–3.

11 See M. V. Ramana, *The Power of Promise: Examining Nuclear Energy in India* (New Delhi: Penguin India, 2012), and references therein.

12 Andrew Lichterman, "Political Will: Civil Society, Social Movements, and Disarmament in the 21st Century," in Ray Acheson, ed., *Assuring Destruction Forever: Nuclear Weapon Modernization Around the World* (New York: Reaching Critical Will, 2012), 135.

13 Johanna Höffken and M. V. Ramana, "Nuclear Power and Environmental Injustice," *WIREs Energy and Environment* (September 21, 2023): e498, and references therein.

14 M. V. Ramana, "Technical and Social Problems of Nuclear Waste," *WIREs Energy and Environment* 7, no. 4 (July/August 2018): e289.

15 Suvrat Raju and M. V. Ramana, "The Other Side of Nuclear Liability," *Economic and Political Weekly* 45, no. 16 (2010): 48–54; M. V. Ramana and Suvrat Raju, "The Impasse over Liability Clause in Indo-U.S. Nuclear Deal," *India Ink* (*New York Times* blog), October 15, 2013, india.blogs.nytimes.com; Athena Kerins and M. V. Ramana, "Liability for Nuclear Accidents: Whose Interests Are Served?," in K. Sudha, ed., *Conserve for Future*, Environmental Law Series II (Visakhapatnam, India: Damodaram Sanjivayya National Law University, 2021), 1–24.

16 I am aware that "developing countries" is a problematic term. Alternative terms like "Global South" or "Third World" are also problematic in other ways. I use the term "developing countries" simply because that is the most common term used in mainstream discussions of nuclear energy.

1. Undesirable

1 Thomas Bass, "Made in Japan," *Mekong Review*, July 2020, 17.

2 National Academies of Sciences, Engineering, and Medicine, "John Downer presentation – What can be learned from other industries," September 2021, video, August 1, 2022, youtube.com.

3 Charles Perrow, *Normal Accidents: Living with High-Risk*

Technologies, rev. ed. (Princeton, NJ: Princeton University Press, 1999), vii.

4 Ali Ahmad, Andrei Covatariu, and M. V. Ramana, "A Stormy Future? Financial Impact of Climate Change–Related Disruptions on Nuclear Power Plant Owners," *Utilities Policy* 81 (April 1, 2023): 101484.

5 M. V. Ramana and Ashwin Kumar, "'One in Infinity': Failing to Learn from Accidents and Implications for Nuclear Safety in India," *Journal of Risk Research* 17, no. 1 (January 2, 2014): 23–42, and references therein.

6 Amy King and M. V. Ramana, "The China Syndrome? Nuclear Power Growth and Safety after Fukushima," *Asian Perspective* 39, no. 4 (2015): 607–36, journals.rienner.com.

7 Avino Niphi and M. V. Ramana, "Talking Points: Narrative Strategies to Promote Nuclear Power in Turkey," in Majia Nadesan, Martin Pasqualetti, and Jennifer Keahey, eds., *Energy Democracies for Sustainable Futures* (London: Elsevier, 2022).

8 Ashwin Kumar and M. V. Ramana, "Compromising Safety: Design Choices and Severe Accident Possibilities in India's Prototype Fast Breeder Reactor," *Science and Global Security* 16 (2008): 87–114, tandfonline.com; Ashwin Kumar and M. V. Ramana, "The Limits of Safety Analysis: Severe Nuclear Accident Possibilities at the PFBR," *Economic and Political Weekly* 46, no. 43 (October 22, 2011): 44–9; M. V. Ramana and Ashwin K. Seshadri, "Negligence, Capture, and Dependence: Safety Regulation of the Design of India's Prototype Fast Breeder Reactor," *Journal of Risk Research* 18, no. 8 (2015): 1030–50.

9 David Lochbaum et al., *Fukushima: The Story of a Nuclear Disaster* (New York: New Press, 2014), 13.

10 Benjamin K. Sovacool et al., "Comment on 'Prevented Mortality and Greenhouse Gas Emissions from Historical and Projected Nuclear Power,'" *Environmental Science and Technology* 47, no. 12 (June 18, 2013): 6715–17.

11 Brice Smith, *Insurmountable Risks: The Dangers of Using Nuclear Power to Combat Global Climate Change* (Takoma Park, MD: IEER Press, 2006); M. V. Ramana and Ashwin Kumar, "Safety First? Kaiga and Other Nuclear Stories," *Economic and Political Weekly* 45, no. 7 (2010): 47–54; David Lochbaum, "Nuclear Power in the Future: Risks of a Lifetime," *Bulletin of the Atomic Scientists*, February 24, 2016, thebulletin.org; Spencer Wheatley, Benjamin Sovacool, and Didier Sornette, "Of Disasters and Dragon Kings: A Statistical Analysis of Nuclear Power Incidents and Accidents," *Risk Analysis* 37, no. 1 (January 1, 2017): 99–115.

12 M. V. Ramana, "Technical and Social Problems of Nuclear Waste," *WIREs Energy and Environment* 7, no. 4 (July/August 2018): e289; M. V. Ramana, "Nuclear Power and the Public," *Bulletin of the Atomic Scientists* 67, no. 4 (2011): 43–51.

13 For this analogy, I am drawing on a conversation with Gordon Edwards (The Canadian Coalition for Nuclear Responsibility) and K. S. Shrader-Frechette, *Burying Uncertainty: Risk and the Case against Geological Disposal of Nuclear Waste* (Berkeley: University of California Press, 1993), 2.

2. Infeasible

1 This picture is distorted by one country: China. Over this period, China connected forty-seven reactors to its grid but closed no reactors. In 2022, nuclear energy contributed around 5 percent of electricity generated in China; wind and solar energy, together, contributed nearly thrice as much electricity as nuclear plants.

2 See M. V. Ramana, "Second Life or Half-Life? The Contested Future of Nuclear Power and Its Potential Role in a Sustainable Energy Transition," in Thijs Van de Graaf et al., eds., *The Palgrave Handbook of the International Political Economy of Energy*, Part IV: "Energy Transitions," ed. Florian Kern (London: Palgrave Macmillan, 2016), and references therein.

3 I use "source of power" in a general sense and am including batteries and demand response within this ambit. For more, see Amory B. Lovins and M. V. Ramana, "Three Myths about Renewable Energy and the Grid, Debunked," *Yale E360*, December 9, 2021, e360.yale .edu.

4 This is the figure from the BEIS 2016 report for the levelized cost of nuclear power, adjusting for inflation. It did not update its cost estimate in the 2020 report—presumably because there was no new data, in turn because there were no new projects.

5 In the case of solar energy, the development phase takes under five years.

6 Bernadette K. Cogswell et al., *Nuclear Power and Small Modular Reactors in Indonesia: Potential and Challenges* (Jakarta, Indonesia, and Berkeley, CA: Indonesian Institute for Energy Economics and Nautilus Institute for Security and Sustainability, April 2017), nautilus.org, and references therein.

7 M. V. Ramana and Priscilla Agyapong, "Thinking Big? Ghana, Small Reactors, and Nuclear Power," *Energy Research and Social Science* 21 (November 2016): 101–13, and references therein.

8 For example, C. G. Suits, the vice president and director of research at the General Electric Corporation, declared in 1950: "At present, atomic power presents an exceptionally costly and inconvenient means of obtaining energy, which can be extracted more economically from conventional fuels ... The economics of atomic power are not attractive at present, nor are they likely to be for a long time in the future." See C. Guy Suits, *Suits: Speaking of Research* (New York: John Wiley, 1965), 28.

9 So much so that three academics from the University of Pennsylvania argued that despite sunk costs, it would be financially more profitable to just stop constructing several of these reactors. See Stephen L. Feldman, Mark A. Bernstein, and Robert B. Noland, "The Costs of Completing Unfinished US Nuclear Power Plants," *Energy Policy* 16, no. 3 (June 1, 1988): 270–9.

10 The term "nuclear renaissance" itself was much older, having been put to use in the 1980s after the cancellation of nuclear power plant orders. See Edward J. Markey and Douglas C. Waller, *Nuclear Peril: The Politics of Proliferation* (Cambridge, MA: Ballinger Publishing, 1982); Joseph F. Pilat, Robert E. Pendley, and Charles K. Ebinger, eds., *Atoms for Peace: An Analysis after Thirty Years* (Boulder, CO: Westview Press, 1985); Alvin M. Weinberg et al., "The Second Nuclear Era: A Nuclear Renaissance," *Energy* 10, no. 5 (May 1, 1985): 661–80.

11 Jennifer Wilson, "Gimmicks Might Be the Key to Understanding Capitalism: Interview with Sianne Ngai," *Nation*, June 11, 2020.

12 The question also reveals the perspective that many nuclear advocates have of countries in Africa, or other developing countries for that matter, seeing them as corrupt guinea pigs for Western technological experiments. Offering further evidence for that perception is Kugelmass's response to a GE official's statement about the need for any vendor to deal with local and regional politics: "In Uganda that's like a five-million-dollar problem. Hire the ... head of government, you hire his son to be your chief lobbyist, and five million bucks you got the whole things solved ... a five-million-dollar marketing campaign making Ugandans love nuclear so much that it's not even an issue."

13 M. V. Ramana and Ashwin K. Seshadri, "Negligence, Capture, and Dependence: Safety Regulation of the Design of India's Prototype Fast Breeder Reactor," *Journal of Risk Research* 18, no. 8 (2015): 1030–50; Frank von Hippel, "It Could Happen Here," *New York Times*, March 23, 2011.

14 Daniel Ford, *The Cult of the Atom: The Secret Papers of the Atomic Energy Commission* (New York: Simon and Schuster, 1982), 194.

3. Private Profits, Social Costs

1 A holding company is a commonly used corporate scheme to reduce liabilities to business owners. According to Wikipedia, the name SCANA was not an acronym, but was taken from the letters in "South Carolina" (S [outh]-C-A-[roli] N-A).

2 Nuclear Energy Institute, "Steve Byrne of SCE&G on How Paying for Nuclear Construction Costs before Operation Saves Money," video, July 23, 2012, youtube.com.

3 Cassandra Jeffery and M. V. Ramana, "Big Money, Nuclear Subsidies, and Systemic Corruption," *Bulletin of the Atomic Scientists*, February

12, 2021, thebulletin.org. There is also widespread corruption of the more common kind in the nuclear industry. One academic study that examined just "the eighteen months from the beginning of 2012 to mid-2013" found evidence of "major corruption incidents" in the "nuclear power industry in every country currently seeking to export nuclear reactors: the United States, Canada, Japan, South Korea, Russia, France, and China" as well as in other countries, like "Lithuania, Bulgaria, and Pakistan." See Richard Tanter, "After Fukushima: A Survey of Corruption in the Global Nuclear Power Industry," *Asian Perspective* 37, no. 4 (October 1, 2013): 478. A further illustration of standard corruption is offered in Shakiba Fadaie and M. V. Ramana, "A Dirty Battle for a Nuclear Bailout in Ohio," *Bulletin of the Atomic Scientists*, April 21, 2020, thebulletin .org. For a more up-to-date review of corruption and criminality, see Mycle Schneider and Antony Froggatt, *The World Nuclear Industry Status Report 2021* (Paris: Mycle Schneider Consulting, September 2021), 218–34, worldnuclearreport.org.

4 Tom Henry, "Global Debate over Nuclear Power's Role in Reducing Greenhouse Gases Intensifies," *Blade*, January 26, 2022, toledoblade .com.

5 "Entergy Agrees to Post-Shutdown Sale of Pilgrim, Palisades Nuclear Power Plants to Holtec International for Decommissioning," Entergy Corporation, August 1, 2018, entergynewsroom.com; Bruce Gellerman, "Pilgrim's Progress: The Pace of Decommissioning Plymouth's Nuclear Plant Picks Up," WBUR, September 16, 2021, wbur.org.

6 Holtec has already received funding from the US government to build a facility to store spent fuel at the site of the Chernobyl nuclear power plant in Ukraine—presumably for the geopolitical goal of diverting payments away from Russia, the alternative. WNN, "Ukraine Secures US Funding for Storage Facility," *World Nuclear News*, February 15, 2018, world-nuclear-news.org. The facility was inaugurated in November 2021 and preparing to receive fuel from other nuclear plants in Ukraine when Vladimir Putin launched his brutal attack on Ukraine, with Russian soldiers taking over the Chernobyl site for a few weeks.

7 *The World Nuclear Waste Report 2019: Focus Europe* (Berlin and Brussels: World Nuclear Waste Report, 2019), worldnuclearwaste report.org.

8 Isha Bhasin, Sara Nelson, and M. V. Ramana, "How Electric Utilities Thwart Climate Action: Politics and Power," *Against the Current*, February 2022; Cathy Kunkel and M. V. Ramana, "Holding Clean Energy Hostage," *Jacobin*, August 23, 2016, jacobin.com.

9 Examples include Pillsbury and King & Spalding; see Jacqueline Thomsen, "Pillsbury Firm Ends Decade-Long Work for Saudi Arabia on Nuclear Energy," Reuters, September 2, 2022, reuters.com.

4. Enabling Moneymaking, Singing Praise

1 Alex Callinicos, *New Labour or Socialism?* (London: Bookmarks Publications, 1996), marxists.org. Blair's specific target was the constitution's pledge "to secure for the producers by hand and brain the full fruits of their industry, and the most equitable distribution thereof that may be possible, upon the basis of common ownership of the means of production, distribution, and exchange and the best obtainable system of popular ownership and control of each industry or service."

2 Stephanie Cooke, *In Mortal Hands: A Cautionary History of the Nuclear Age* (New York: Bloomsbury USA, 2009), 351; see also Peter Wilsher, Donald Macintyre, and Michael Jones, *Strike: Thatcher, Scargill and the Miners* (London: A. Deutsch, 1985).

3 Stephen Thomas, "Is New Nuclear Build Justified?" (Press conference on the need for an independent inquiry, Westminster, March 11, 2010), nuclearconsult.com.

4 Back in 2007, de Rivaz had pronounced that the lights would go out in the UK by 2017 without Hinkley Point. Emily Gosden, "Has Nuclear Been Stuffed by Hinkley Turkey?," *The Times*, December 23, 2017.

5 Mycle Schneider and Antony Froggatt, *The World Nuclear Industry Status Report 2013* (Paris: Mycle Schneider Consulting, 2013), 112.

6 There are two reasons why the example of Pakistan is special. First, China shares a special relationship with Pakistan and has provided the latter with nuclear weapons and related military technology. Second, because of Pakistan's nuclear weapons program, no other country has been willing to export nuclear reactors to Pakistan since the 1970s. Given its relative weakness in nuclear technological capability, Pakistan has little choice but to import from China.

7 See Amy King and M. V. Ramana, "The China Syndrome? Nuclear Power Growth and Safety after Fukushima," *Asian Perspective* 39, no. 4 (2015): 607–36, and references therein.

8 To understand the background to this decision, see Sara Nelson and M. V. Ramana, "Managing Decline: Devaluation and Just Transition at Diablo Canyon Nuclear Power Plant," *Environment and Planning A: Economy and Space* 55, no. 8 (2023): 1951–69.

9 Mark Hand, "Kochs Invest in Company Set to Benefit from Coal and Nuclear Bailout," *ThinkProgress* (blog), June 12, 2018, thinkprogress.org. Although his fortunes were made in the oil and gas industry, Charles Koch has a master's degree in nuclear engineering from the Massachusetts Institute of Technology and does not have any ideological opposition to the technology. See Jane Mayer, *Dark Money: The Hidden History of the Billionaires behind the Rise of the Radical Right* (New York: Doubleday, 2016), 50–4.

10 Cassandra Jeffery and M. V. Ramana, "Big Money, Nuclear Subsidies,

and Systemic Corruption," *Bulletin of the Atomic Scientists*, February 12, 2021, thebulletin.org.

11 There may be occasions when the benefits to the larger public justify subsidies, but clearly nuclear power does not fit that qualification.

12 Shakiba Fadaie and M. V. Ramana, "A Dirty Battle for a Nuclear Bailout in Ohio," *Bulletin of the Atomic Scientists*, April 21, 2020, thebulletin.org.

13 M. V. Ramana and Eri Saikawa, "Choosing a Standard Reactor: International Competition and Domestic Politics in Chinese Nuclear Policy," *Energy* 36, no. 12 (December 2011): 6779–89; M. V. Ramana and Zia Mian, "Scrambling to Sell a Nuclear Middle East," *Bulletin of the Atomic Scientists* 72, no. 1 (2016): 39–43.

14 Rose Gottemoeller, "Geopolitics and Nuclear Energy: The View from the State Department," US Department of State, May 15, 2013, state .gov.

15 Stephen Thomas and M. V. Ramana, "A Hopeless Pursuit? National Efforts to Promote Small Modular Nuclear Reactors and Revive Nuclear Power," *WIREs Energy and Environment* 11, no. 4 (2022): e429.

16 Mycle Schneider and Antony Froggatt, *The World Nuclear Industry Status Report 2023* (Paris: Mycle Schneider Consulting, 2023), worldnuclearreport.org.

17 See Bernadette K. Cogswell et al., *Nuclear Power and Small Modular Reactors in Indonesia: Potential and Challenges* (Jakarta, Indonesia, and Berkeley, CA: Indonesian Institute for Energy Economics and Nautilus Institute for Security and Sustainability, April 2017), nautilus.org, and references therein.

18 See M. V. Ramana and Priscilla Agyapong, "Thinking Big? Ghana, Small Reactors, and Nuclear Power," *Energy Research and Social Science* 21 (November 2016): 101–13; Avino Niphi and M. V. Ramana, "Talking Points: Narrative Strategies to Promote Nuclear Power in Turkey," in Majia Nadesan, Martin Pasqualetti, and Jennifer Keahey, eds., *Energy Democracies for Sustainable Futures* (London: Elsevier, 2022), and references therein.

19 Noam Chomsky, "'Mandate for Change,' or Business as Usual," *Z Magazine*, February 1993, chomsky.info.

20 Some of the solar- and wind-energy-related jobs are not unionized or are relatively low paying. For people on the left, this situation should serve as a call to unionize these sectors and push for better quality jobs.

21 For example, I have calculated that just paying the operators of a hypothetical Oklo nuclear power plant would make the cost of electricity nearly three times that of new solar or wind power plants, even if it costs nothing to build the reactor or fuel it. See M. V. Ramana, "Small Modular and Advanced Nuclear Reactors: A Reality Check," *IEEE Access* 9 (2021): 42090–9.

22 For more on the role of the Justice and Development Party (AKP) in furthering Turkey's nuclear program, see Niphi and Ramana, "Talking Points," and references therein.

5. May the Atom Be a Soldier

1 Harold A. Feiveson, "In Memoriam—Ted Taylor," *Science and Global Security* 13, no. 1–2 (January 1, 2005): 117–28.
2 Harold A. Feiveson et al., *Unmaking the Bomb: A Fissile Material Approach to Nuclear Disarmament and Nonproliferation* (Cambridge, MA: MIT Press, 2014), 43–67.
3 "Saudi Crown Prince: If Iran Develops Nuclear Bomb, So Will We," CBS News, March 15, 2018, cbsnews.com.
4 French nuclear experts were among the most vociferous in maintaining this position, in part because of France's large investments in reprocessing of spent nuclear fuel. See Frank von Hippel, Masafumi Takubo, and Jungmin Kang, *Plutonium: How Nuclear Power's Dream Fuel Became a Nightmare* (Singapore: Springer, 2019), 65.
5 Quoted in George Perkovich, "Nuclear Power and Nuclear Weapons in India, Pakistan, and Iran," in Paul L. Leventhal, Sharon Tanzer, and Steven Dolley, eds., *Nuclear Power and the Spread of Nuclear Weapons: Can We Have One without the Other?* (Washington, DC: Brassey's, 2002), 194.
6 Zia Mian, "Fevered with Dreams of the Future: The Coming of the Atomic Age to Pakistan," in Itty Abraham, ed., *Nuclear Power and Atomic Publics: Society and Culture in India and Pakistan* (Bloomington: Indiana University Press, 2009), 35.
7 Robert Anderson, *Building Scientific Institutions in India: Bhabha and Saha* (Montreal: Center for Developing Area Studies, 1975), 101.
8 Aileen Murphy and M. V. Ramana, "The Trump Administration Is Eager to Sell Nuclear Reactors to Saudi Arabia. But Why?," *Bulletin of the Atomic Scientists*, April 16, 2019, thebulletin.org.
9 M. V. Ramana, *The Power of Promise: Examining Nuclear Energy in India* (New Delhi: Penguin India, 2012).
10 Bechtel and AECOM are not alone; General Electric, for example, was found to have overcharged the navy for nuclear submarine propulsion equipment. GAO, "Annual Report of the Comptroller General of the United States" (Washington, DC: US Government Printing Office, 1968), 160, gao.gov; US Government Accountability Office, "General Electric Company Claim Shown in GAO Report to Congress," May 18, 1971, gao.gov.
11 The third member of Reagan's "Bechtel Cabinet" was Defense Secretary Caspar Weinberger, formerly the firm's general counsel.
12 US Department of Energy, addendum draft, May 29, 2018, available as "Grid Memo," contributed by Bloomberg News, at https://www.documentcloud.org/documents/4491203-Grid-Memo.html.

13 Members included, among others, Maria Korsnick, president and chief executive officer of the Nuclear Energy Institute; Peter Lyons, former assistant secretary for nuclear energy, US Department of Energy; Thomas Graham, a former ambassador and the executive chair of a nuclear fuel supply company called Lightbridge Corporation; Richard Meserve and Jeffrey Merrifield, both former commissioners at the US Nuclear Regulatory Commission; and Admiral Richard Mies, fourth commander in chief of the United States Strategic Command from 1998 to 2001.

14 Theodore B. Taylor, "Nuclear Power and Nuclear Weapons," *Nuclear Age Peace Foundation* (blog), July 12, 1996, wagingpeace.org.

6. Magical Thinking and Billionaire Messiahs

1 M. V. Ramana and Zia Mian, "One Size Doesn't Fit All: Social Priorities and Technical Conflicts for Small Modular Reactors," *Energy Research and Social Science* 2 (June 2014): 115–24.

2 See M. V. Ramana and Cassandra Jeffery, "Bill Gates and Techno-Fix Delusions," *Against the Current*, October 2022, againstthecurrent. org, and references therein.

3 M. V. Ramana, *Eyes Wide Shut: Problems with the Utah Associated Municipal Power Systems Proposal to Construct NuScale Small Modular Nuclear Reactors* (Portland: Oregon Physicians for Social Responsibility, September 2020), oregonpsr.org, and references therein.

4 Arjun Makhijani and M. V. Ramana, *Questions for NuScale VOYGR Reactor Certification: When Will It Be Done? And Then, Will It Be Safe?* (Environmental Working Group, 2023), ewg.org, and references therein.

5 Makhijani and Ramana, *Questions for NuScale*; Ramana, *Eyes Wide Shut*.

6 Alexander Glaser et al., *Small Modular Reactors: A Window on Nuclear Energy. An Energy Technology Distillate* (Princeton, NJ: Andlinger Center for Energy and the Environment at Princeton University, June 2015), acee.princeton.edu.

7 Ibid.

8 Arjun Makhijani and M. V. Ramana, "Can Small Modular Reactors Help Mitigate Climate Change?," *Bulletin of the Atomic Scientists* 77, no. 4 (2021): 207–14.

9 M. V. Ramana, "The Forgotten History of Small Nuclear Reactors," *IEEE Spectrum*, May 2015, spectrum.ieee.org.

10 M. V. Ramana, "The Checkered Operational History of High-Temperature Gas-Cooled Reactors," *Bulletin of the Atomic Scientists* 72, no. 3 (April 2016): 171–9, and references therein.

11 Matthias Englert, Friederike Frieß, and M. V. Ramana, "Accident Scenarios Involving Pebble Bed High Temperature Reactors," *Science*

and Global Security 25, no. 1 (January 2017): 42–55, and references therein.

12   M. V. Ramana, "Even China Cannot Rescue Nuclear Power from Its Woes," Center for Asian Studies Brief, April 12, 2022, colorado.edu.

13   S. Rajendran Pillai and M. V. Ramana, "Breeder Reactors: A Possible Connection between Metal Corrosion and Sodium Leaks," *Bulletin of the Atomic Scientists* 70, no. 3 (2014): 49–55.

14   Thomas B. Cochran et al., *Fast Breeder Reactor Programs: History and Status* (Princeton, NJ: International Panel on Fissile Materials, 2010), fissilematerials.org.

15   M. V. Ramana, "Molten Salt Reactors Were Trouble in the 1960s— and They Remain Trouble Today," *Bulletin of the Atomic Scientists*, June 20, 2022, thebulletin.org, and references therein.

16   Ramana and Mian, "One Size Doesn't Fit All."

17   Jungmin Kang and M. V. Ramana, "Canadian Reactors That 'Recycle' Plutonium Would Create More Problems Than They Solve," *Bulletin of the Atomic Scientists*, May 25, 2023, thebulletin.org.

18   Benjamin K. Sovacool and M. V. Ramana, "Back to the Future: Small Modular Reactors, Nuclear Fantasies, and Symbolic Convergence," *Science, Technology, and Human Values* 40, no. 1 (2015): 96–125; on early visions of an atomic utopia, see Paul Boyer, *By the Bomb's Early Light: American Thought and Culture at the Dawn of the Atomic Age* (Chapel Hill: University of North Carolina Press, 1994).

19   Edwin Lyman, *Small Isn't Always Beautiful: Safety, Security, and Cost Concerns about Small Modular Reactors* (Cambridge, MA: Union of Concerned Scientists, September 2013); Ahmed Abdulla, Inês Lima Azevedo, and M. Granger Morgan, "Expert Assessments of the Cost of Light Water Small Modular Reactors," *Proceedings of the National Academy of Sciences* 110, no. 24 (2013): 9686–91; M. V. Ramana and Priscilla Agyapong, "Thinking Big? Ghana, Small Reactors, and Nuclear Power," *Energy Research and Social Science* 21 (November 2016): 101–13.

20   Kerrie Blaise and M. V. Ramana, "Regulation vs. Promotion: Small Modular Nuclear Reactors in Canada," *Energy Policy*, submitted (n.d.).

Conclusion

1   Karl Marx, *Capital*, vol. 1 (London: Penguin, 1990), 493, footnote 4.

2   Sharachchandra Lele, "Environment and Well-Being," *New Left Review*, no. 123 (June 22, 2020): 41–63, newleftreview.org; Lindah Ddamba, Sara Nelson, and M. V. Ramana, "Energy, Injustice and the Green New Deal," *Progressive International*, August 25, 2020, progressive.international.

3   James Baldwin, "As Much Truth as One Can Bear," *New York Times*, January 14, 1962.

Index